Beamtimes and Lifetimes

BEAMTIMES AND LIFETIMES

The World of High Energy Physicists

SHARON TRAWEEK

Harvard University Press
Cambridge, Massachusetts
London, England
1988

Printed in the United States of America
10 9 8 7 6 5 4 3 2 1

This book is printed on acid-free paper, and its binding materials have been chosen for strength and durability.

Library of Congress Cataloging in Publication Data

Traweek, Sharon.
Beamtimes and lifetimes : the world of high energy physicists / Sharon Traweek
p. cm.
Includes bibliographical references and index.
1. Physicists. 2. Particles (Nuclear physics)—Laboratories—Environmental aspects. I. Title.
QC774.A2T73 1988 88-11035
306'.45—dc19 CIP
ISBN 0-674-06347-3 (alk. paper)

I dedicate this book to my niece, Victoria Ashley Traweek, my young cousins Michelle and Valerie Nichols, my godson, Hosaki Fujita Sano, and my young friends Mary and Arthur Richardson. I also dedicate it to the memory of two teachers, the late Gregory Bateson and the late Michelle Zimbalist Rosaldo.

Contents

Preface

Ethnographies traditionally open with an introductory chapter that, in addition to stating the topic, surveys the academic literature, identifies unresolved problems, defines research questions, explicates methods, and proposes hypotheses, and the chapter is written, as far as possible, in the passive voice. Studies of the rhetorical forms adopted in science writing have shown the extraordinary stability of these forms through three centuries and how they are reinforced by faculty, journals, and funding agencies. In recent years, however, some anthropologists have been moving away from writing narrative accounts of fieldwork in a form and style in which the author claims to be remote, authoritative, definitive.

Anthropology first emerged in collaboration with colonialism and missionary activities, and the traditional style of writing an ethnography supported the objectives of these endeavors: classification and control. As those objectives are abandoned by anthropologists, new modes of inquiry have emerged in which questions of the nature of knowing, of the relation of knower and known, and of how that relationship shapes understanding are no longer taken for granted. From this perspective it is crucial for the anthropologists to write themselves into their accounts and to describe the conditions under which they formed their interpretations. As anthropologists move away from writing definitive accounts of other cultures and toward writing accounts of local strategies for making sense—that is, accounts of how the world is interpreted in cultures—the modes of presenting our own interpretations are changing. As part of that shift I have chosen a form for this book that underscores the interaction between me, the physicists I study, and the readers of this book. Overall, the book has four parts: the prologue introduces the actors, including myself, and there I address the readers directly. The first three chapters present in detail the spaces, artifacts, and roles of the actors in the community; the last two chapters describe the activities that are important in the lives of physicists and of national physics communities; and the epilogue frames my reflections on what *knowing* means for the physicists.

The first chapter explores laboratory spaces, the environments the high energy physics community builds for itself. Touring a laboratory, from the imposing office buildings to the catacombs holding the accelerator, the labyrinthine switchyards bending the accelerated particles to their targets, and the stark sheds in the research yards, is a way of mapping the community, its divisions and relations, of seeing who uses and controls each of these spaces. I was amazed during my first visit to the Stanford Linear Accelerator Center; the bizarre shapes and juxtapositions of the accelerator and research yard reminded me of the strangely beautiful industrial landscapes of Antonioni's film, *Red Desert*. As I gradually learned to read the spaces at that lab and others, I realized that almost imperceptibly the lab sites elicited certain forms of action and discouraged others.

The second chapter introduces and describes some detectors, the devices whose design is at the heart of the experimental process. I discuss what counts to the physicists as interesting differences in the design and operation of these massive research devices. Like the environments we build, the artifacts we make remind us of who we are, how we are expected to act, and what we value. Physicists fashion detectors to record some very elusive traces of "natural" events at the subatomic level, events which would be completely unobservable without this apparatus. The detector is a research group's means of production of knowledge, the tool the group builds and uses to make its livelihood. To the experimentalists each detector has its physiognomy, full of meaning and value. In the features of a detector we can learn to read a group's history, its division of labor, its strategy for discovery.

The third chapter tells how physicists are made. The high energy physics community reproduces itself by winning new recruits. Novices must learn what sorts of things they need to know to be taken seriously; they must become unselfconscious practitioners of the culture, feeling the appropriate desires and anxieties, thinking about the world in a characteristic way. They learn, in succession, the distinctive qualities befitting each stage of a career, the marks of success or failure for undergraduate, graduate student, postdoctoral research associate, group member, group leader, laboratory director, and science statesman.

The fourth chapter describes some of the stable features in the way physicists act toward each other. The community has a relatively fixed and highly articulated hierarchical structure: countries, laboratories, universities, and research specialities are ranked with surprisingly slight disagreement and with little change over decades. The community is divided into theorists and experimentalists—moieties which are interdependent, each with distinctive duties. Cross-cutting these distinctions are international, overlapping networks of exchange that hold the community together. Physicists seem to be in constant circulation, moving around the world from lab to lab, department to department, always talking, forming alliances and collaborations. Most important, they are bound together by a way of thinking, about the world and about knowledge and about themselves.

The fifth chapter is about change. All physicists try to keep themselves, their ideas, and their equipment at "the cutting edge," the "frontier of knowledge," the "state of the art." To do this they need to gain priority in access to accelerators, access to the best recruits, access to funding. They negotiate with one another for these resources, they forge factions, they enter into and resolve disputes. I describe the strategies that physicists in the United States and Japan have used to build and maintain the highest quality in research equipment at their respective labs.

The epilogue focuses on the relation between the physicists' *theories* of time and their *experience* of time in their working lives. High energy physicists learn about conflicting kinds of time through their work: the negotiable and cumulative "beamtime"—pulses of the accelerator beam—and the intractable and limited lifetimes of their careers, their detectors, and their ideas. The products of their work, however, are perceived as timeless: the unchanging laws of nature. I conclude my account with a discussion of how the tension inherent in these attitudes and actions is reflected in the social world of high energy physics.

The primary fieldwork for my study was done in the mid-1970s, but my description uses a standard convention, the "ethnographic present." It is not that the world of high energy physics has remained unchanged in the last ten years; far from it. The detectors, the accelerators, the directors of laboratories, the funding have all changed. Everyone has aged; there are new novices, and new ideas. The world in which the physics community is embedded has changed. Yet it is striking that the physicists' underlying ways of constructing detectors, novices, leaders, and laboratories, their ways of making sense and of making money, have changed very little. Partly this is because the generation that launched high energy physics in its current form at the end of World War II has remained in charge for forty years; only now are the founding fathers beginning to retire. The same generation has led in industry, government, labor, and academia; their web of personal relations has made for great overall stability. It is now, in the late 1980s, that we are beginning to witness the first major transfer of leadership in forty years. What I have sketched here is the world the new physics leaders inherit, with its current models of tradition and change.

Acknowledgments

First, I would like to express my deep gratitude to all the high energy physicists, their families, and their staffs who have given me the opportunity to do fieldwork among them; their overwhelming intellectual and personal generosity is what makes my work possible. The ethics of anthropology require that I not identify my "native informants" nor the specific events and actors I describe; consequently, the only physicists I name in this book are public figures.

Jehane Kuhn of Boston kindly consented to edit this book. The word *edit* does not properly describe her contributions. Before addressing the manuscript itself she made many inquiries about my goals for this book and my approach to anthropology, fieldwork, and the high energy physics community. At every stage of our work together she kept these issues at the center of our discussion, asking of each sentence, each paragraph, each page exactly how it engaged those issues. Nearly every page of what follows shows traces of her fine hand and I am very grateful for it. Howard Boyer, Science Editor at Harvard University Press, not only suggested I work with Jehane, he facilitated every other phase of the production of this book and I am much indebted to him for his unending good taste and kindness. Kate Schmit, Editor at Harvard University Press, persisted wonderfully in the final stages of making this book presentable; she found and tied every loose thread.

I know that the warp and woof of my thoughts have been strung by my teachers. My undergraduate teachers at the University of California at Berkeley have no reason to remember me, but their books and lectures continue to be immensely important for me; they include Carlo Cipolla, Lewis Feuer, Paul Feyerabend, Robert Paxton, John Schaar, Carl Schorske, Raymond Sontag, George Stocking, and Sheldon Wolin. In graduate school, first at the San Francisco State University History Department and later at the University of California at Santa Cruz Program in the History of Consciousness, I worked very closely with my teachers; as I weave the patterns and textures of my ideas, I still feel them guiding my hands. These patient tutors include the late Gregory Bateson, James Clifford, Shelly Errington, Vartan Gregorian (now at the New York Public Library), Triloki Nath Pandey, Thomas P. Rohlen (now at Stanford University), the late Michelle Zimbalist Rosaldo, and Hayden White.

In 1980 I began teaching for the Stanford University Program in Values, Technology, Science, and Society; in 1981 I joined the MIT Program in Anthropology and Archeology *and* the Program in Science, Technology, and Society; and in 1987 I moved to the Rice University Anthropology Department. The continuing generous guidance and kind support of several fine mentors have sustained me during my peregrinations through the world of American academia. They include, in addition to the above, James Adams, Edwin Good, and Walter Vincenti (all of Stanford), L. L. Bucciarelli, Martin Diskin, James Howe, Jean Jackson, Carl Kaysen, Kenneth Keniston, Thomas Kuhn, Heather Lechtman, Leo Marx, Ruth Perry, Merritt Roe Smith, Benson Snyder, Arthur Steinberg, Sherry Turkle, Charles Weiner, and Joseph Weizenbaum (all of MIT), and Michael Fischer, George Marcus, and Stephen Tyler (all of Rice). Helene Moglen (UC Santa Cruz), Amelie Rorty (Radcliffe), and Laura Nader (UC Berkeley) have provided absolutely crucial *extra mural* guidance.

When I was just finishing graduate school I met some established researchers in the social studies of science and in gender and science studies who very generously began to include me in their ongoing conversations. Their professional generosity has been extremely important to me; they are Karin Knorr, Bruno Latour, Sal Restivo, and Steve Woolgar in science studies and Donna Haraway, Sandra Harding, and Evelyn Fox Keller in gender and science studies. My work would be thoroughly impoverished without theirs, which has established and continues to define entire fields of research.

While teaching at MIT I met some young researchers whose own fine work has strongly influenced mine: they are Craig Decker, Roz Gerstein, Scott Globus, Brenda Golianu, Vera Ketelboeter, Akiko Kodaka, Gloria Lee, R. Ruth Linden, Scott Minneman, Monica Strauss, and Ramon Solorzano. Their intellectual independence has been a joy to me.

The exciting work and kind words of my friends Barton Bernstein, Frank Dubinskas, Anna Hargreaves, Billie Harris, Chela Sandoval, Susan Foster, Mariko Fujita, Katie King, Martin Krieger, Marlys Macken, Wendy Martin, and Judith and Philip Richardson have inspired and sustained me for many years. Everyone in the commune at 540 Seale gave me much needed encouragement when I returned to school in my mid-thirties. The accomplishments achieved against incredible odds by my mother, Edna Traweek, my brother, Richard Traweek, my cousin, Mary Williams Nichols, and my father, S. V. Traweek, have inspired and sustained me as long as I have known them. The kindness of my uncle, Cecil Lewis, and my sister-in-law, Victoria Goetz Traweek, has been wonderful.

Earlier versions of this book were returned to me, heavily annotated or with essays appended, by several kind and thoughtful readers among those already mentioned. In addition I want to thank John Pfeiffer, David Schneider, and Roy Wagner for their lucid commentaries. The misconceptions and errors in the text are most certainly mine, but I do hope that this book somehow honors those who have, however unwittingly, collaborated in its conception, gestation, and birth. Unfortunately, these acknowledgments do not properly convey the extraordinary and fundamental contributions to my work by those people. The book in your hands is a palimpsest, my words, under erasure, written over by the first readers, followed by my jottings over their notes made as I read through their readings.

Beamtimes and Lifetimes

PROLOGUE

An Anthropologist Studies Physicists

The public and political role of high energy physicists, the role of science as a secular religion, as well as the history, sociology, and philosophy of physics are the subjects of many important books by eminent scholars. This book is not about how physicists have shaped our world, or why our society has given them power and prestige. Nor is it about the current state of knowledge and inquiry in high energy physics. Instead it is an account of how high energy physicists see their own world; how they have forged a research community for themselves, how they turn novices into physicists, and how their community works to produce knowledge.

Since World War II physicists have maintained a special hold on the American imagination. Their discoveries are front-page news; *Time* magazine tries to describe the latest theoretical developments. Autobiographies of leading physicists reach best-seller lists. The image of Albert Einstein is still used as an emblem of intelligence and creativity. When Prime Minister Fukuda of Japan and President Carter met during the energy crisis of the 1970s, they decided to fund a ten-year, multimillion-dollar research project in high energy physics. During the budget-cutting years of recession all the major industrial countries continued to increase the funding of the enormously costly research in this field; and in 1987 President Reagan announced his support for a new "superconducting supercollider" many times larger than the largest high energy physics facility ever before built.[1] How is it that physics and physicists have so strong a claim on our society?

Part of the answer is war. Competition for novel weapons during World War I led to the organization of research and development labs staffed by scientists and engineers. The modest successes of these labs brought funding during World War II to diverse and esoteric projects, and this time the yield was much greater: it included radar and, most decisively, the atomic bomb. The search for the fundamental secrets of nature was suddenly a matter of national power and prestige. The bomb brought political influence to the "atom smashers," and with that influence came money; neither has declined in the succeeding forty years, in spite of the fact that since the war almost all high energy physicists have refused to do secret research or to work on weapons.[2]

Part of the answer is organization. In the new mission-oriented labs of World War II, high energy physicists learned to administer large interdisciplinary teams of researchers, manage huge budgets, and speak the language of government agencies. At the end of the war, many of these same scientists were called in to help reorganize and redefine the goals of the agencies, which have funded basic research ever since; high energy physicists have maintained personal ties and influence in Washington. At the same time, their organizational skills and political acumen have not gone unnoticed in the universities: the expansion of the resources of physics departments is the envy of other disciplines, and many senior high energy physicists have become university deans, provosts, and presidents.[3]

Yet another part of the answer is the emotional power of cosmology. The physicists' calling is awesome: memoirs and biographies often present this corps d'elite as unique, Promethean heroes of the search for truth. Traditionally the mysteries of the universe have been the province of theologians and priests. Physicists of course do not see themselves as writing the cosmology of some secular religion: for them, religion is about belief rather than knowledge. But they do see their own profession as the revelation and custody of fundamental truth, and to a surprising degree Western culture confirms them in this privileged role. They bring news of another world: hidden but stable, coherent, and incorruptible. In times of bewildering and threatening change, this gospel, however esoteric, has a very deep appeal. (Fear may confuse these feelings, but it does not weaken them.) The extraordinary scale and costliness of much physics research if anything reinforces its cultural value. The great accelerators, for example, are like medieval cathedrals: free from the constraints of cost-benefit analysis.

The physicists who work in these great accelerators study the basic constituents of matter and the elementary forces that operate between them. This field of study is called particle physics.[4] (Nuclear physics is a separate subject area, dealing not with the components of the nucleus, but with the relations between nuclei. Its implications lie mainly in nuclear energy, material science, and medicine.)

Though strong in influence, high energy physicists are not strong in number. According to the international leaders of the community, there are about eight hundred to a thousand very active researchers in the world in their field. They suggest that perhaps two thousand more are abreast of the latest developments. Three or four hundred of all these people know one another quite well, and all the other practitioners want to.[5] About half of the group are theorists; they work at blackboards alone or in small, short-lived collaborations of two or three people. The other half are experimentalists; they work with big machines in long-lived groups of twenty to fifty people. The experimentalists and theorists need the others' contributions to solve their scientific problems, but they keep a predictable, friendly distance from each other, and they are readily distinguishable at any laboratory by their styles and habits. In this book I will be writing primarily about experimentalists.

High energy physicists gather together to do research at five major accelerators, in Western Europe, the USSR, the United States, and now Japan. The first major American accelerator was constructed at the University of California at Berkeley in the 1930s under the leadership of Ernest Orlando Lawrence, who was indefatigable in his search for money to support the lab. His accelerator received financial support from several sources: Lawrence publicized the medical applications of procedures developed for basic research in order to gain funding from an eclectic collection of individuals, governmental agencies, private business, and philanthropic foundations. He claimed that the knowledge gained by the new machine was the "beginning of an economic revolution"; when his staff created gold from platinum Lawrence announced that "the information we are getting is worth more than gold."[6] His public appeal that physics should be funded because it ultimately enhances the public good has been copied by today's physicists.

Lawrence's role in the design, construction, research, and funding of the accelerator at Berkeley established a style in laboratory research that is maintained now throughout the particle physics community. The close bonds between a laboratory, its research, and its director is found in nineteenth-century science; Lawrence's innovation was a shift in scale and an adroit use of public opinion to gain funding.

Funding for the high energy physics community is directly determined by national governments: no private sponsorship could maintain a field dependent on machines so massive and so constantly changing. Science funding in both Japan and the United States is determined first by the national government as part of annual budgets. Before this budget with its science research component is voted upon, particle physicists gather into a national high energy physics advisory panel (HEPAP) to determine which projects they will advocate, and hence what allocations they will seek, as a concerted community. Construction of new facilities and operating budgets for the laboratories are established at this level. This committee reports to the Department of Energy (DOE) and the National Science Foundation (NSF) in the United States, and to the Ministry of Education (Monbusho) in Japan. Membership on this committee is an honor and a position of great power. Once priorities have been established, various members begin to advocate the projects and their funding to appropriate government agencies and legislative committees; certain individuals are known to be particularly effective at this crucial stage. A few physicists regularly encourage some of their students to become science advisors in Washington so that they will have loyal representatives working with key committees.

Each laboratory in turn has its own program advisory committee (PAC), which, in principle, sets long-range science policy for the lab and decides which proposed experiments will be done and how much accelerator time will in principle be allocated to each experiment. All accepted experiments are funded entirely from the laboratories' overall budgets. Established groups within the labs also receive funding from the lab budgets for maintenance and development of their research equipment, known generically as "detectors." Access to the accelerator and a detector is crucial in experimental particle physics, and it is the lab's program advisory committee that decides which groups will have that access.

Once experiments are accepted by the program advisory committee they are scheduled by the "long-term scheduling committee," which determines how to coordinate the various experiments simultaneously at the laboratory. Groups want a certain number of pulses per second from the accelerator beam over a number of months ("beamtime") directed to their detector; in addition, they also want scheduled access to computer time and other laboratory facilities. The long-term scheduling committee has the power to allocate time within the PAC's guidelines. As the slate of experiments draws near, the "short-term scheduling committee," another powerful group, assumes control over distribution of time, including reallotments of accelerator beamtime if they are needed because of accelerator or detector failure. The "beam switchyard," which delivers accelerator beam pulses to the diverse detectors, becomes the final arbitrator in last-minute revisions of time distributions. Conflict at this stage is intense. Access to beamtime is a precondition of power in experimental particle physics.

The community I have just described is scarcely the sort of group that is generally thought of as material for anthropological study. Classical anthropologists studied small communities that were non-Western, nonindustrialized, nonliterate . . . in short, as the string of negatives suggests, not like us. The methods and theories that evolved in such situations have in fact been applied to "our" world—to the sort of "developed" societies of which anthropologists themselves are a part—for at least fifty years. A notable early example is Ruth Benedict's *Chrysanthemum and the Sword*, a study of Japanese values undertaken during World War II to assess the role of the emperor in Japanese society. But such studies have been marginal to the main enterprise of anthropology until quite recently.[7]

The new, "repatriated" anthropologists study people with power as well as those without, corporations as well as ghettoes.[8] The status of the inquiring anthropologist changes accordingly, from a knowing, benevolent visitor, backed by technical and political power, to a tolerated, perhaps amusing, marginal presence. These days "informants" offer employment to anthropologists, not the other way around. The physicists I study are always shocked at the low pay of anthropologists and our small research grants; they regularly offer advice on "using the system" more effectively. Most significantly, we have become informants ourselves; our reports and interpretations are read and debated in the communities we study. The anthropologist no longer has the last word in the dialogue of fieldwork: when I submit a manuscript for publication, those who are asked to review its merits always include physicists. And since an anthropologist often studies the same community throughout her entire career, the conversation between us may go on for forty years.

Anthropologists study relatively small communities, usually not larger than three to five thousand people. A community is a group of people who have a shared past, hope to have a shared future, have some means of acquiring new members, and have some means of recognizing and maintaining differences between themselves and other communities. The high energy physics community meets this definition. The first condition of anthropological study is that the anthropologist live in a community long enough to observe a full cycle of routine activities. In an agricultural society, it takes the four seasons of a year to observe all the activities associated with food production. In a research laboratory, a full cycle comprises the planning and execution of an experiment.

Completing an experiment in high energy physics can take three to five years. First, a research group selects a physics question of current interest to the community and decides on a procedure for confirming one of the proposed answers. It then designs equipment for carrying out this procedure; as I shall discuss in more detail later, high energy physics experiments almost always depend on innovations in equipment. The entire experiment, with costs, is described in a written proposal, typically of about 150 pages, which is submitted to the lab's Program Advisory Committee (PAC). At the same time group members give lectures at other laboratories and university physics departments, describing the virtues of their proposal.

If the PAC accepts the proposal, the research group begins "tooling up," drafting detailed descriptions of the equipment and software and assigning primary responsibility to individuals for various components of the experiment. Production and assembly will usually take one to three years. Conducting the experiment itself may take about one year; during that time the group monitors, repairs, and modifies the equipment and begins to analyze the data being collected. The group may revise the equipment, the software, or the shape of the experiment, if early data suggest that a change is needed. After the experiment is completed, a year or so of intensive data analysis begins. When the group members agree that they have found interesting data, they arrange to give another round of lectures, trying to persuade their colleagues of the significance of their results. If their arguments are well received, the speaker and other members of the group "write up their data" in articles submitted to scientific journals. All the members of the group, in alphabetical order, are listed as authors, although informally everyone comes to know who contributed what. Copies of submitted articles are circulated as "preprints"; as a rule active researchers read preprints, not journals. Graduate students write up for their doctoral thesis their own contribution to the experimental design, to the equipment, or to the analysis, but theses are seldom read except by the candidate's teachers. The important communications are made by word of mouth: by informal talk and by lectures and seminars. In five years of fieldwork I witnessed each of these stages of producing an experiment.

The account written as an outcome of anthropological fieldwork, an ethnography, usually includes information about four domains of community life.[9] The first is ecology: the group's means of subsistence, the environment that supports it, the tools and other artifacts used in getting a living from the environment. The second is social organization: how the group structures itself, formally and informally, in order to do work, to form factions, to maintain and resolve conflicts, and to exchange goods and information. The third is the developmental cycle: how the group transmits to novices the skills, values, and knowledge that constitute a sensible, competent person; the stages of a life and the characteristic attributes of a person at each of those stages. The fourth is cosmology: the group's system of knowledge, skills, and beliefs, what is valued and what is denigrated.

These four domains can be separated only in an artificial way, for purposes of analysis: in all human action all four are present, in some configuration distinctive to the group. Any ethnography, whatever its primary focus, must address all four domains and the relations between them if it is to contribute to an account of the culture of the group. One current definition, developed in the work of David Schneider and Clifford Geertz, makes "culture" a group's shared set of *meanings*, its implicit and explicit messages, encoded in social action, about how to interpret experience.[10] The ethnographer tells how those meanings are generated, maintained, and transmitted in different ecological settings, and how they affect the group's ecology. An ethnography describes *patterns* of explanation and action, the meanings people bring from one situation to another, the connections and distinctions people make between certain actions, feelings, ideas, things, and their environment; these patterns make up the culture.

The fieldworker's goal, then, is to find out what the community takes to be knowledge, sensible action, and morality, as well as how its members account for unpredictable information, disturbing actions, and troubling motives. In my fieldwork I wanted to discover the physicists' "common sense" world view, what everyone in the community knows, and what every newcomer needs to learn in order to act in a sensible way, in order to be taken seriously. I wanted to understand how young physicists in Japan and the United States learn the ethos of their community, how they are immediately recognizable to others as high energy physicists; how the young person comes to display good physics judgment, commitment and trustworthiness; how physicists, research equipment, and data are taken to be reliable and trustworthy; why these qualities are so valued in the community; and what happens to those who violate these expectations. The first goal is to discover what counts as being the right kind of person in the community one studies.

I wanted to find out how the physicists generate the shared ground that all members of the community stand upon; how they define the established terrain within which debate can occur, the recognized strategies for making data and equipment and reputations, and the ground rules for contesting data, machines and reputations. Describing and explaining how knowledge in science and technology is *contested* is the subject of many books and articles published during the past fifteen years. Through meticulous case studies researchers have shown how scientists and engineers use their accumulated resources (reputation, funding, students, technicians, and laboratory space, equipment, and techniques) to make strategic experimental choices, how they make and make use of data to construct and defend their intellectual positions, how they recruit supporters and defeat critics, and how the written accounts of their work reflect and conceal this elaborate and stylized combat. I have chosen to describe how scientists and engineers construct the ground on which this contest is waged, how they all can agree on what can be contested, how they all can agree on what is an interesting or a boring contest. I believe that to understand how scientific and technological knowledge is produced we must understand what is *uncontested* as well as what is contested, how the ground state is constructed as well as how the signals called data are produced. When I speak of the shared ground I do not mean some a priori norms or values but the daily production and reproduction of what is to be shared.[11] In my research I wanted to find the forces of stability, the varieties of tradition, in a community dedicated to innovation and discovery.

Anthropologists collect life histories, stories about knowledge, legends, myths, and theology, as well as information about networks, generational relations, negotiation, leadership and followership, conflict, change, and stability; they also observe the construction of artifacts ("material culture") and collect descriptions of them. This kind of fieldwork, conducted in settings arranged and structured by the community, not by the researcher, is known as participant-observation. The "observation" aspect calls for detailed attention to how the people in the community conduct themselves in daily life: the goal is a "thick description" of settings, language, tone of voice, posture, gestures, clothing, distance, arrangement of movable objects, and how all this changes from one interaction to another. The "participant" aspect calls for the fieldworker to take account of how the group responds to her, the stages by which she gradually comes to be accepted or at any rate tolerated.

The anthropological literature is rife with "fables of rapport," feelings of communion with the subjects of investigation. Whether we like or dislike our informants, whether they like or dislike us, does not determine the validity of our observations or interpretations. A good fieldworker explores the foundations of empathy and antipathy and uses them to examine the local commonsense notions of what counts as the "obvious" responses, the adult way to act in everyday interactions. She explores the local social construction of emotion, of gender, how people come to want to do what they should, and how they cope with anomalies and transgressions.

The fieldworker needs to remain marginal. If she were to become a fully integrated participant in the community, its sociocultural assumptions would no longer stand out in the foreground of her attention; and in any case it would no longer then be appropriate for her to be asking questions about the meaning of social actions. On the other hand, if she "learns" too slowly, her informants may become exasperated, impatient, or bored. It is part of the job to make social "mistakes," to note which of her actions are accepted as worthy, which are treated as inappropriate, repulsive, or ridiculous, and how approval or disapproval is conveyed. All this is taken as information about how the group maintains its boundaries and guides its own members toward acceptable behavior. The fieldworker learns to ask if her own embarrassment and anxiety are culturally induced emotions, designed to make one wish to do what one should do in a given culture.

Clearly, inquiry by participant-observation cannot maintain the distinction between subjective and objective knowledge. It does not assume that the relations between investigator and subject should be distant or dispassionate, still less that the investigator should control that relationship or the subject. To the contrary, it requires the investigator to have a close and complex relation to the subject, and to be rigorously conscious of her "objective" and "subjective" understanding of the community as well as the interaction between her observations and her affective responses. Physicists, of course, adhere to a strikingly different model of inquiry—which has led to many, many discussions between us about the nature of knowing and knowers.

The major fact of this kind of research is that the fieldworker lives her days and weeks and months within the patterns of the community's life, moving in spaces shaped by the community and taking part in its activities on its terms. As a fieldworker she learns what the informants take to be interesting, boring, useful, catastrophic, funny, fortunate, troubling, exciting; she learns the right actions for each situation, how things fall apart and how they are mended. While participating she must try to observe and remember every detail. This intense awareness is quite difficult to sustain, in part because she receives no direct appreciation for it. The lack of reinforcement and the inability to lead her life on her own terms are the fieldworker's burden. At the beginning she chafes; toward the end she feels torn from the life she has learned to lead. This separation is rehearsed daily as she retreats to write up the day's and night's observations. It is from these field notes, from the records of interviews, from collected documents and artifacts, and from the experience etched in her memory that the process of analysis and the process of writing the ethnography begins.

My fieldwork was conducted at three national laboratories over a period of five years: National Laboratory for High Energy Physics (Ko-Enerugie butsurigaku Kenkyusho, or KEK) at Tsukuba, Japan, Stanford Linear Accelerator (SLAC) near San Francisco, and Fermi National Accelerator Laboratory (Fermilab) near Chicago. I also visited other laboratories, including CERN in Geneva and DESY near Hamburg, and several university physics departments. My inquiry began several years before my formal commitment to it as an anthropological study. In 1972, while I was a graduate student in intellectual history, I took a part-time job in SLAC's Public Information Office. Along with three or four other graduate students, I explained the activities of the lab to visitors, who ranged from junior high school to college students, from the general public to special interest groups like safety experts, electrical engineers, and chemists, from new employees to visiting dignitaries. There was no training for this job: we were simply told to start asking questions, to find the right people to talk to. After about a month of learning how to learn, I began work. Over the next three years, I came to know a lot about the lab, and from seminars and conferences in the auditorium where I ran the sound system I learned more about physics. I also read textbooks and autobiographies, as well as the history and philosophy and sociology of physics. Increasingly, I took every opportunity to talk with people informally about the lab, and I began to take my own curiosity more seriously in the context of my own discipline of intellectual history.

My graduate research topic had been a study of social and technological change in the French Protestant textile industry in the early nineteenth century. Gradually it became clear that I wanted to change it: I planned to write a history of SLAC. As word of this spread around the lab, people who already knew me would stop me in hallways to tell me "important information" for my research; some began to show me files of memos, notes, and reports they had been saving. I was delighted by the promise of all these documents.

As I learned more stories about the past, I became fascinated by the ways they conflicted with one another. As a graduate student I knew that it was part of the historian's task to find the truth among these conflicting stories: which of them were "correct"? But I soon recognized that this was not where my own interest lay: what I wanted to know was why these conflicting versions had survived into the present, told by people who saw one another daily. As I studied the methods and models of oral history, I discovered that anthropologists were concerned with explaining why communities maintained competing versions of the past in the present. In 1975 I made the change to anthropology. Gregory Bateson, my dissertation advisor at the outset, advised me to go to Japan so that the American labs would become strange to me, to sharpen my observations of them. At SLAC I had met Japanese physicists during conferences and they had urged me to visit; I made the arrangements and went there in the spring of 1976, officially beginning my anthropological fieldwork.

It can be difficult to gain access to a community as a fieldworker. The group can be suspicious of the researcher's motives, skeptical of her intellectual or social or emotional capacities. Even if access is granted, informants may withhold real rapport. Inquiry can be stifled with politeness and formality; a fieldworker's endless questions about what seems obvious, natural, mere common sense, can be tedious or unnerving, since she is dwelling on the unexamined assumptions of the community. Since the people of the laboratory I set out to study already knew me well, the story of my fieldwork experience does not hinge on gaining access, but on the way the shift in my roles, when I reappeared at SLAC as an anthropologist, was managed by the group.

As a Public Information Officer I stood outside social divisions and yet was a familiar part of the lab. I knew and talked with many people in different parts of the lab, because I had to find out about their current work in order to give the tours. I was not associated with any particular group, as I had soon realized that in order to do my job well I needed to subtly dissociate myself from my supervisors in the Public Information Office, which was seen as tainted by undue preoccupation with the outside world. People joked that they did not recognize me without the rickety old bus in which I ferried visitors around the laboratory grounds. In the auditorium during seminars and lectures I was seen hooking up sound systems, fixing broken equipment, showing slides, and listening attentively. Versatility, especially around mechanical equipment, is a recommendation among experimentalists. Before I decided on the shift to anthropology, I already knew that people were willing to talk to me freely. As I described to some of the senior scientists my growing interest in studying the high energy physics community, they offered encouragement; they told me of the issues that concerned them, that they would like to see studied. When the time came, some wrote letters of support for my project, helping me to make the transition to the new discipline.

In a sense, then, when my fieldwork began there were already figures standing ready to act as "key informants." For the anthropologist, key informants are crucial; they are people with whom one can try out tentative interpretations and hypotheses. People who are interested in consciously reflecting on their own culture tend to be atypical within it, whether leaders, geniuses, or simply marginal; they are willing to reflect on the differences between themselves and their fellows, with amusement, sympathy, bitterness, remorse, detachment, or condescension. (Ultimately, the anthropologist should be able to account for why certain kinds of action lead to those roles at the edge of a community.) Key informants are indispensable; but it is essential for the fieldworker not to become too closely identified with any specific informant or faction in the eyes of the community. The danger is that one's work will become an account of a particular viewpoint and of the community's reactions to that viewpoint.

As my fieldwork went on I noticed sharp differences in the responses of senior physicists and of their most junior colleagues. The senior people, once they had scheduled an interview with me, stopped phone calls and closed their office doors; anyone interrupting was asked to return later. They invariably expressed interest in my study, gave thoughtful responses to my questions, and often asked me how others were responding. Each of them seemed to feel responsible for setting me straight and helping me get the right picture; they were clearly disappointed when they felt I failed to understand that I was "wrong." My first task was to suppress my own conditioned impulse to show myself a good student—which they elicited so well. They liked my persistent inquisitiveness, although they did not like me to pursue questions they had decided were unimportant. Some of these men had experience of interviews with journalists and historians and questionnaires from sociologists. Occasionally they would discuss the similarities and differences between my approach and that of others.

By contrast, when I interviewed the youngest physicists in their offices very few of them closed their doors. Passersby would pause and stare, and often interrupt to carry on a conversation with my interviewee, who would break off our exchange without ceremony. The counterpart of this disregard was very valuable; they were much more unguarded in their responses than their seniors; indeed, they often spoke to me as if they were free-associating. It was as if—being both a nonphysicist and a woman—I was not in a position to use special knowledge to gain power. Once I asked what kind of information was generally considered worth keeping secret. "Well," my informant answered, "one thing we never tell anyone is . . ." It was clearly to my advantage that I was not anyone, but it was sometimes hard to take. I had to dissemble my anger, to resist devaluing myself and my work, and to take note as an ethnographer of the tremendous force of the division in the physicists' cosmology between outsiders, no matter how well-informed, and insiders. In 1987, when I am noticeably older than the junior physicists I interview, they still respond in the same way, in sharp distinction to the courtesy of the senior and midcareer scientists, who are both more secure in their own positions and more alert to the uses of the outside world.

Having done fieldwork in Japan during 1976 and again ten years later, I have found that there, too, prominent senior physicists are much more accessible and reflective about their community than their junior and less established colleagues. But there is a major difference: Japanese scientists in general are very interested in the differences between one national scientific community and another, not only in the organization and funding of research, but also in culture. Japanese physicists are quite conscious of belonging to an emerging scientific community, still building the so-called infrastructure of basic research: sound scientific education and modern laboratory equipment at all levels of the school system, excellent research laboratories at universities, state-of-the-art research equipment for basic researchers, an emerging "critical mass" of first-class scientists, and sustained funding to maintain all this. They know that they need these resources in place if they are to participate as equals in the international high energy physics community, and they are eager to learn how American physicists have gained and protected the social resources that have enabled them to build their own world-class laboratories and staff them with first-rate scientists. They want to know how Americans make decisions among themselves about the allocation of these resources. They are interested in what parts of the American practice of science are distinctively American, and these are my concerns too. Hence, Japanese curiosity about my observations of the American community has often been the starting point of our conversations.

There are other differences also. At a Japanese laboratory there is no way that a five-foot-eight green-eyed auburn-haired woman can expect to fade into the background; and my activities are almost as strange as my appearance. Until quite recently there has been little public interest in high energy physics in Japan, and very little sociology of science; the idea of being interviewed and studied is new and surprising to Japanese scientists. I am always observed at least as much as observing.

It might seem that an American fieldworker would share the commonsense cultural expectations of American physicists. Traditional anthropologists have sometimes been suspicious of "repatriated" anthropology, arguing that without strangeness the fieldworker cannot identify the cultural assumptions of a community; they believe that shared common sense is transparent and hence invisible.

The premise is surely right, but the cases where it applies are rarer than they may seem. For one thing, repatriated anthropology is just that; one of the main reasons I worked in Japanese labs before returning to SLAC was to acquire strangeness. And it is important to remember that there is plenty of strangeness within the United States; in spite of the image of the melting pot, regional, class, ethnic, religious, and occupational differences give rise to sharply differing experiences of the world. In fact, many features of the Japanese physicists' social interactions—their expectations for men and women, their sense of family, what is said and not said, and their ways of expressing joy and anger, frustration, and relief—seem to me very familiar, evocative of the way I myself was raised by my traditional Southern family.

As I began to construct an account of physicists' culture, certain landmarks emerged. Three in particular persisted as key *symbols* of the culture: the physicists' experiences of time, the artifacts called detectors, and a way of thinking that is sometimes called "realism." I have used these symbols in mapping the field of acceptable strategies for making sense and being successful in the community.

There are, in fact, three intersecting cultures involved here: that of the international physics community, and those of Japan and the United States. In the context of their intersections, I have focused on two activities in particular: the training of novice physicists and the management of changes in the structure of laboratories. Both the training of novices and the renewal of institutions are examples of, and models for, maintaining tradition and achieving change—key problems in cultures dedicated to innovation.

It is during apprenticeship that traditions of research are passed on. Patterns of apprenticeship differ in United States physics and in Japanese physics, in ways connected to the larger national cultures in which they are embedded. Institutional change also, and resistance to it, follows different but related patterns in Japan and the United States. The two particle physics labs on which my study is concentrated, KEK and SLAC, face quite similar institutional challenges. Both continue to be confronted with the problem of balancing the interests of resident research groups—"insiders"—with those of visiting experimentalists, usually university based, known as "users." These interests concern decisions about, access to, and management of experimental facilities at each laboratory. Senior physicists are forced to reevaluate what is the best organizational environment for physics.

Two recurring themes in my account are gender and national culture. In the fifteen years I have been visiting physics labs, the status of women within them has remained unchanged—in spite of major transformations, in North America and Europe, in opportunities for women and attitudes about their roles. In this book, women remain marginal, as they are in the laboratory. The lab is a man's world, and I try to show why that is particularly the case in high energy physics: how the practice of physics is engendered, how laboratory work is masculinized.

Among national physics communities, the Japanese have also been peripheral. But they are moving away from their position on the sidelines, and their presence in this book reflects that. Until recently Japan sent money and its best researchers abroad to bring data home; now it is beginning to bring talented researchers and students from around the world to Japan. Japan may soon rank among the world centers of basic scientific research; and high energy physicists are at the vanguard of this movement, struggling to maintain momentum. Their efforts to establish KEK, and to bring its equipment to world standards and keep it there, are part of the larger goal of moving Japan from the periphery to the core of the international community. I have tried to understand how their strategies for this campaign are related to Japanese research traditions.

Deeper even than submerged assumptions about gender and national identity are profound and deeply felt tensions about time that I find coiled at the center of this culture. In the course of a career a physicist learns the insignificance of the past, the fear of having too little time in the present, and anxiety about obsolescence in the face of a too rapidly advancing future. But the *content* of high energy physics, its explanations of the interactions of fundamental particles, screen out all consideration of elapsing, ephemeral time. Although the physicists speak of generations of particles, of lifetimes and decay, they represent the natural world as a stylized dance of only a few forces, endlessly repeating a small set of choreographed interactions. They have a passionate dedication to this vision of unchanging order: they are convinced that the deepest truths must be static, independent of human frailty and hubris. Simultaneously, they believe that this grand structure of physical truth can be progressively uncovered, and that this is the highest and most urgent human pursuit. Their everyday anxieties about the terrible loss of time—terrors that are carefully maintained in the culture of physics, as if they were essential driving forces for the good physicists—seem to me a mirror image of the cosmological vision that transcends change and mortality. I came to this view by spending many hours and months around detectors, coming to see them as embodying all their builders' divergent meanings and experiences of time. The detectors in the end are the key informants of this study; physicist and nature meet in the detector, where knowledge and passion are one.

CHAPTER · 1

Touring the Site: Powerful Places in the Laboratory

Almost all of the high energy physics national laboratories in the United States, and many of those around the world, have lovely settings. The Stanford Linear Accelerator Center rests unobtrusively in the scenic eastern foothills of the Santa Cruz Mountains, thirty miles south of San Francisco near Stanford University. During the winter rainy season, the rolling grasslands are a bright malachite green; gnarled California live oaks stand isolated, forming black veins in the grey skies and green savannah. Sharp air and fulminous clouds frame Mount Diablo, far across the San Francisco Bay. In the thick, dry sunlight of summer, slowly moving shadows from the great oaks trace small arcs in the still, golden grasses. The weather of this region alternates between a five-month-long winter of rain and a seven-month-long summer of sun. The accelerator at the laboratory operates about seven months of the year. Just as the ecology of alternating wet and dry seasons shapes the social and political institutions of many nonindustrial, nonurban peoples, the ecology of an alternating "on" and "off" accelerator shapes the social organization of the laboratory.[1]

Several herds of cattle graze on the 480-acre site, and many other animals live there, secluded and protected in Jasper Ridge Biological Preserve and the Stanford Outdoor Primate Facility, where research is conducted on plants and animals in a controlled simulation of their natural habitats. The twenty-four-acre Stanford Outdoor Primate Facility (SOPF) was founded in 1974 by David Hamburg at Stanford University and Jane Goodall, who also founded the Gombe Stream Research Center in Tanzania, East Africa, where chimpanzees are studied in their natural habitat. Certain kinds of experimental studies cannot be conducted in fully natural settings; the controlled but varied environment at the SOPF suited the research conducted by the Laboratory of Stress and Conflict of the Department of Psychiatry and Behavioral Sciences at the Stanford School of Medicine.[2]

Particle physics research is also conducted in two settings: one uses the naturally occurring cosmic rays which are constantly bombarding the earth, and the other uses artificially accelerated particles. Like SOPF, SLAC offers greater control of research variables than any natural setting could provide.

Although the San Andreas earthquake fault runs visibly alongside the site about half a mile from the accelerator, the bedrock and earthmass of the site have been stable for ten to fifty million years. Measured to a depth of thirty feet over eighteen months, the maximum horizontal shift was a quarter-inch north to a quarter-inch south, and the total vertical movement was less than an eighth-inch.[3] The stability of the site is necessary to assure the alignment of the accelerator.

Approaching the lab by car from Sand Hill Road, one is aware only of the bucolic landscape and a small, low grey concrete marker lettered in white: "STANFORD LINEAR ACCELERATOR CENTER: Operated for the United States Department of Energy." No one at the lab calls it that; it is known there and elsewhere in the physics community as SLAC (pronounced "slack"). It is common in the high energy physics community to devise acronyms with humorous or mundane associations.[4]

For visitors turning at the sign, the first impression is of a small building large enough for two or three people. It is referred to as an information booth; it looks like a guardhouse because the middle-aged men who stand there watching the cars come and go wear uniforms and some occasionally execute sharp salutes, apparently for their own amusement. The men work for a private security firm that has a contract from the lab. Passersby who notice the guardhouse sometimes think that access to the lab is restricted. This impression could be dispelled if the administration chose—for example, by a sign on the booth indicating that the lab is open to the public and giving directions to the Public Information Office.

Some people who work at SLAC object to the quasi-military

demeanor of the guards; others simply joke about the booth and the men in it, saying the whole display looks like a Swiss clock. The style of the guards offends the physicists because it reminds them of the gates at labs where "classified" (secret) research is done on behalf of the military. Secret work is distasteful to them because it is seen as "applied" research, in which ideas already established in "basic" or "pure" research are applied to less fundamental and challenging problems. They are proud of working at a lab where no classified work is conducted, because in their eyes basic research has much higher status.

The guards have various small movable wooden signs. On winter mornings they post the one that says "Your Lights Are ON." Once a month they display "Bookmobile Here Today." At night the sign reads "Stop Here 7 P.M. to 6 A.M." During those hours it is necessary to show a lab identification badge and sign a roster before being allowed to enter the site. No one objects to the procedure; it is regarded as protecting the lab from theft, vandalism, and political terrorism.

Since two bombs exploded at the western end of the accelerator on December 7, 1971, causing several thousand dollars' worth of damage, the administration of SLAC has been especially concerned with the security of the lab.[5] The rule about registering at the booth dates from that incident; and the Klystron Gallery, which gives access to the accelerator, is now kept locked. With the bombing incident the lab's relationship with the outside world became an explicit issue for the first time. No one was ever apprehended; speculation at the lab identified three possible groups that might have been willing to use violence against the institution. Whichever it was, they were seen as misguided and irrational in wishing to disrupt the important work of the lab; the feeling was that, whatever the validity of the grievances of the firebombers, the lab was an inappropriate target.

The first hypothesis was that the firebombing was the work of radical left-wing anti–Vietnam War activists who assumed that SLAC was engaged in classified weapons research. The physicists find annoying the association in the public mind between physics and weapons research.

Their second theory assumed the bombing must have been an "inside job," their suspects being the people who were then trying to establish the first union at a government-financed research laboratory. They were later successful: the United Stanford Employees (USE) won the National Labor Relations Board–sponsored vote over the Teamsters. From the beginning USE leaders have been concentrated at SLAC—particularly among the technicians—and more widely dispersed on the Stanford campus as a whole. The senior physicists who manage the lab are said to have been outraged at the very idea of a union, which suggested that people could consider their work at the lab as a mere job. The physicists generally have been committed to being scientists since early adolescence, and their own training teaches them to regard physics as a calling, not an occupation. They assumed that everyone at the lab would share this same devotion to science and its institutions and were profoundly saddened and angered by the strike and picket lines called by the union in 1973. The lab has generally attracted a staff devoted both to SLAC and its work. They, as well as the physicists, are uncomfortable working with employees who seem not to share this commitment. It is assumed that those who do would not consider a strike against the lab, because it would be a strike against science.

The third hypothesis suggested that the bomb was the work of nearby residents. Many neighbors were frightened of radiation or possible explosions (harking back to the association of physics with weapons research). Others were known to be offended by the huge power lines that provide electricity for the lab: the lines pass through an affluent suburb many of whose residents affect the style of English gentry, including fox hunts. According to laboratory employees, Pete McCloskey, now a former congressman and then a local attorney, negotiated a settlement between the residents and the federal government concerning these power lines. The residents wanted the lines to be placed underground no matter what the expense, but Lyndon Johnson decided against this after his personal emissary, Laurance Rockefeller, informed him that this was not the current community standard. The compromise struck was to have the poles designed by a prestigious San Francisco architectural firm, Halprin Associates, and to keep them painted green. The physicists regarded all these concerns as silly, a sign of ignorance, and a confusion of priorities.

After these early conflicts with the community, the director established a Public Information Office to educate citizens and students about the work conducted at the lab through news releases and tours. The head of this office eventually became a popular city councilman. His demeanor and values are much closer to those of the local community than to those of the scientists at the lab. The citizens consider this man a representative of their point of view to the lab; at SLAC he is regarded as a delegate from the lab to the community. Through him each side feels comfortable communicating with the other.

All of the employees of the Public Information Office are Caucasian, like almost all of the audiences they address. The activities of this office indicate that SLAC sees its role in community-laboratory interactions as didactic. That is, disagreement with the lab's policies is seen as a result of lack of information, which SLAC will supply.

By the rules of the government agency overseeing the laboratory, the Department of Energy (formerly the Atomic Energy Commission, then—at the time of this study—the Energy Research and Development Agency, ERDA, and now the Department of Energy, DOE), the lab cannot openly advertise this community service, because such notices could be construed as using public funds to solicit support for the lab's budget. The ads routinely placed in the Stanford University *Daily* are acceptable within the letter of the law because they are defined as intraorganizational notices.

The shift from passive dispensation of information upon request to more assertive public relations is a move made in the 1970s by SLAC as well as other laboratories. This development is analogous to the changed stance of the "statesmen" of high energy physics in Washington, D.C., and the aggressive political action of laboratory directors on behalf of their laboratories in the last forty years.[6]

Mutual self-interest governs this and other aspects of SLAC-University relations. Once a year, a massive tour of SLAC—twenty to thirty busloads of people—is arranged to impress the families of students as a part of graduation week festivities. Powerful and prestigious visitors to the university are generally given personal tours of SLAC. Ties to Stanford are stressed by the SLAC Public Information Office brochures, and in its fund-raising efforts the Stanford Development Office emphasizes the link to the famous research and Nobel Prize–winning work done at SLAC. The formal affiliation between the two institutions is such that the university has a contract from DOE to administer SLAC. Payroll and employee benefits, for example, are dispensed through Stanford. (This is why the USE was able to install a union at a government facility.)

This pattern of mutual assistance, whereby SLAC is used to enhance the prestige of the university and the university is used by SLAC as an administrative buffer that also disseminates information to citizens, is—so far as I know—unique among high energy physics laboratories. The interaction also has other rewards: physicists at SLAC often say that they have the benefits without the liabilities of a university environment. Stanford's science and engineering programs are as extensive and powerful as those at either Caltech or MIT. Having SLAC on campus enhances Stanford's position in the science and engineering communities, although in practice students at the Stanford Physics Department seldom visit SLAC.

In addition to its explicit public relations work, the Public Information Office conducts ten to twenty tours a week with a stronger educational emphasis. Science teachers at high schools and colleges throughout the San Francisco Bay area routinely arrange for their classes to be given tours. Local community and professional organizations also make use of this service. I worked at the lab for three years giving these tours. Although their purpose is didactic and the intentions of the guides are serious, I found most visitors on these tours arrived wanting to be awed rather than informed.

This demeanor of awe was especially marked among practicing scientists and engineers. Visitors often behaved as though they had been granted a special dispensation to see the inner sanctum of science and its most learned priests. Many were quite bluntly dismayed to find a woman guiding them through the hallowed precincts. One chairman of a leading chemistry department, at the head of a group of about seventy-five academic chemists visiting the lab, approached me and asked where the guide was. When I introduced myself, with barely concealed disdain he said, "Well, if they hired you, I suppose you know your stuff."

Once a visitor passes the information booth and its surrounding shrubs of dark green manzanita, the first buildings come into view. In the center of the U-shaped group is a carefully landscaped approximation of the surrounding natural environment. About a dozen live oaks, sitting in disks of loose gravel, are surrounded by a closely cropped but very dense lawn kept green year-round, underscoring the hard sculptural form of the trees. At noon, if the weather is not too soggy, people play volleyball or football on the lawn. By creating an English green lawn in an environment of golden dry savannahs, the lab demonstrates both the authority of its own vision of nature and its power to commandeer water in a land of recurrent drought.

These messages have endangered the only indigenous element: the oaks are quite sensitive and will die if watered through the summer. Planting the trees in gravel is an attempt to have both the lawn and the trees survive. Elsewhere on the site, trees have had to be protected from people parking under them in the summer; the weight of the cars kills the oaks' root systems. An ecosystem has been altered to create both an eloquent *tableau vivant* and a site for massive human enterprise. At least in this case, these two goals are severely conflicting.

On three sides of this grassy square are five of the more than one hundred structures on the site; the road forms the fourth side. These structures are of one, two, and three storeys, with facades of beige and grey aggregate. The flat roofs have deep overhangs, shadowing the many large windows. More shrub manzanita surrounds the buildings, which are linked by asphalt and concrete pathways. From the square and from each of these buildings, there are striking views of the region, including San Francisco Bay.

On a knoll at the left of the square as one approaches are the cafeteria and auditorium, joined by a wide breezeway. The walls of the cafeteria are mostly glass; the limited wall space is covered with rotating exhibits of artwork done by SLAC employees and a blackboard sometimes used by the physicists during their coffee breaks or after lunch. Outside, away from the square and sheltered by oaks, is a patio with several tables and chairs. Altogether there is seating for about one hundred (SLAC has about twelve hundred employees). The food is rather good for a cafeteria; the head of this service takes great personal care of the entire operation, and she and her staff (which is mostly Chicano) are friendly and skilled.[7]

Many people eat at the cafeteria every day, although a few occasionally walk the mile to nearby restaurants, drive to surrounding towns for lunch, or eat near the coin-operated food-dispensing machines around the lab. Sometimes people picnic on the pastoral site away from the buildings, but this is unusual for the physicists.[8]

The secretaries, librarians, and administrative assistants, all but one of whom are women, typically eat together in small clusters. The senior technicians, administrators, and physicists, most of them men, tend not to mingle across job classifications.

It is easy to distinguish between the groups at the cafeteria. The physicists are dressed most casually, in shirts with rolled sleeves and jeans or nondescript slacks. They disdain any clothing that would distinguish them from each other. The style to which they conform, furthermore, maintains a carefully calibrated distance from fashion, quality, or fit.[9] Their general appearance would not be out of place in a middle-class midwestern suburb on a weekend. I can think of four exceptions among over one hundred physicists. One, who wears atypically tight jeans, is clearly identified as English. Another wears suits and is treated accordingly—like an administrator. The third is given to diverse eccentricities, such as purple shirts; he is thought to be anomalous in many ways. The fourth, whose clothes are neatly pressed, well made, and color-coordinated, is said to be really okay: he only dresses "that way" because his wife buys his clothes.

Engineers and senior technicians seem to affect either a collegiate style (khakis, button-down Oxford shirts, and crew-neck sweaters) or a studiously informal appearance (polyester pants and lightly starched shirts). Administrators wear classic business attire, but leave their jackets in the office. Secretaries, administrative assistants, and the few female administrators dress informally but not casually, in dresses and pantsuits. The clothing of the women physicists generally consists of slacks or jeans with a belt and a shirt. In several years at the lab I have seen only one wearing a skirt, and that was only on one occasion.[10]

Usually the physicists from each experimental research group will walk as a body to the cafeteria and then sit together, pulling a few tables end to end and making room for late arrivals. Sometimes senior physicists from different groups have lunch to discuss lab business. Theorists eat in smaller clusters. While eating, people scan the room frequently, noting who is or is not eating with whom. Almost no one eats alone.

Next to the cafeteria, through the breezeway, is the auditorium. It has a small lobby with models and photographs of research equipment at the laboratory. There is also a small "diorama" depicting a paleoparadoxia in its natural habitat—the SLAC site. The fifteen-million-year-old remains of this animal were exposed during excavations on the construction site. A team of paleontologists, led by the wife of the director, had identified the remains and built a full-scale model of the animal in an office at the lab. Similar skeletal remains have been found in Japan, apparently lending support to the plate tectonics theory, which suggests that Japan and California were once adjacent.[11]

Inside the auditorium itself there is seating for about three hundred. The floor is sloped as in a theater, so that the last row of seats is about twenty feet higher than the front row. There is desk space and a small light in front of each seat. In the "stage" area is a lectern, a viewgraph machine for projection from transparencies, six blackboards, and two screens.

All sorts of groups make use of the auditorium. The tours begin here with an introductory lecture lasting about an hour. At lunchtime social, educational, and professional groups organized by SLAC personnel may meet there. A weekly colloquium in theoretical particle physics is held, and also one for experimentalists; these are in the late afternoon. They are widely attended; the director and his deputy and associate directors are almost always there. In the evening community college courses are held, and local civic groups also use the space. Small workshops or conferences are held in the auditorium every few weeks; these run all day for three to five days. These meetings are for physicists, and usually of interest only to people in one particular speciality (theoretical, experimental, or accelerator physics) or concentrating on one topic (such as synchrotron radiation). Every summer the lab hosts a two-week "summer school" (as do other major labs around the world) for graduate students, "postdocs," and scientists who do not work at the major labs. These people are brought up to date on the most recent experiments and theories.

Once a year, the director holds a series of meetings for "all hands" (all laboratory employees) to discuss the achievements and goals of the lab and its funding status for the coming year.[12] The same talk is given several times; people attend the meeting corresponding to the first letter of their last name.

Every two or three years a meeting is held for "users," the experimentalists from other institutions who have applied successfully to make use of the lab's research facilities. The auditorium is the one space on site widely used by diverse groups, both from within and outside SLAC. The time of day, day of the week, and month of the year it is used distinguish one group from another.[13]

On the main axis of the square sits the Central Laboratory and its annex (referred to as Central Lab). The three-storey building has many windows and a few balconies, which are never used (although those offices with balconies, along with the corner offices, have the highest status). The walkway approaching the Central Lab descends a few steps to the entrance. In the foyer are two vending machines, an unused computer terminal, and a metal bookcase holding outdated periodicals and papers discarded by the library. The floor here and throughout the building is beige linoleum.

To the right are the offices of the Technical Illustration Department, which is headed by a woman. The half-dozen employees meticulously transform the physicists' handwritten graphs, equations, and diagrams into a standard format, ready for inclusion in journal articles or transfer to transparencies and slides to be used in oral presentations. Physicists bring their own sketches to the supervisor and discuss with her how they want them prepared and when they will be finished. The technical illustrators are always very busy, and usually each new order must be added to a waiting list. The main room is quite large, has windows on two sides, and is filled with several professional drafting tables. Illustration is one of the few jobs at the lab that are done by people of both sexes. All the illustrators at the time of this study were either Causasian or Asian.

To the left of the foyer is the L-shaped "Orange Room," with seating for about fifty people. Seminars for the experimentalists are held here, as are meetings of laboratory groups such as the SLAC Women's Organization (SLACWO). SLACWO was founded by two senior women employees at the lab for the exchange of information of special interest to women; about 15 percent of the lab employees are women; 25 percent are "minorities." There are seminars and lectures on topics such as educational opportunities for women in business, employee benefits, and statistics on the employment status of women compared to men in the lab, and in the community as a whole. Some women worry that the organization might be identified as feminist or union oriented. Others openly endorse this aspect of the group. The lab administration reluctantly accepted the constitution of the organization, but insists that no laboratory time be devoted to its activities. It is one of the few groups at the laboratory that includes members and officers from all occupational status levels, from physicists to file clerks.

In 1978, one out of 222 female employees (0.45 percent) and eighty-one out of 1,082 male employees (7.49 percent) were regularly engaged in physics research at SLAC.[14] One hundred seventy-seven women at the lab were employed in nontechnical, nonscientific positions. Because of the job classification system in effect at SLAC, there are few opportunities for significant promotions, no matter how skilled and effective one becomes as a manager or administrator; in almost all cases, the higher positions are technical or scientific. The federal agency that oversees all high energy physics laboratories in the United States has required an affirmative action program at each laboratory since early 1975.[15] In my discussions with two affirmative action officers (one at SLAC and one at DOE regional headquarters in Oakland, California), I was told that working with the physicists who ran the labs was extremely frustrating because the physicists seemed not to believe that they were obliged to change their existing hiring practices in any way.

Also on the first floor of the Central Lab are the scanners' rooms. Some of the data from experiments is collected on film. For example, a two-month-long chamber experiment, making three pictures of each event with as many as forty-five events per second, will produce about 775 million frames. Each frame is visually scanned for both predicted and unpredicted information. The scanners work in a large darkroom at big metal tables, with bulky projecting equipment overhead. They use special mechanical tracking arms to follow the traces on the film; the information picked up by these arms is then recorded digitally by computers. The room is silent save for the low whirring of the film tapes being advanced, one frame at a time, and the intermittent clicking of the tracking arms.

Across the hall is the scanners' lounge; they take frequent breaks to relax their eyes. Nine of the thirteen scanners at SLAC are women, some young, some middle-aged; six are black and one is Asian-American. In the 1950s, the early days of bubble chamber physics, film was scanned by young physicists; later the chore was handled by graduate and undergraduate physics students, then by the general student population (this last group did the job as recently as the late 1960s). It is now realized that one need not know anything about physics to do the work well, and its status has declined accordingly.

Behind the scanners' rooms are the offices of two groups of experimentalists, responsible for maintaining and developing certain complex research equipment. In return, these physicists are entitled to do a limited amount of research. Among them are some former students and employees of the full-time research groups. On the rest of the first floor are small workshops for the machining and fabrication of minor equipment needed for experiments; here too are the offices of visiting users, experimentalists from other institutions who are currently making use of the lab's research facilities. Commingled with the users are the offices of the "liaison" physicists: former students, postdocs, or temporary employees whose intimate knowledge of the lab enables them to coordinate the needs of a visiting user group day by day with those of the "in-house" groups, whether for beamtime, computer time, or special machining requirements. At the time of this study, almost all of the approximately one hundred experimentalists at SLAC, including users, were Caucasian (five were Asian, one a Native American); five were women, one of whom was a liaison.

The second floor of the Central Lab is reached by two elevators and a number of narrow, echoing stairwells with concrete stairs and handrails of metal pipe. On the second floor are the windowed offices of experimentalists. The windowless offices in the core of the second floor are occupied by clerks, secretaries, and administrative assistants, all of whom are women; one is black. There are a few exceptions to this arrangement; two administrative assistants to high-status research group leaders have offices with windows; in the Annex, some space at the core is given over to computer terminals and to small electronics workshops.

The offices of the experimentalists are furnished in grey metal: desks, chairs, bookcases (usually filled with computer printouts), and computer terminals. Most offices are shared by two people and feel quite small; both desks and people face the walls. Occasionally one sees a poster of some scenic wilderness area, especially mountains. Doors are rarely, if ever, closed, unless the office is empty.

The offices of the research group leaders are two or three times as large as those of their staff. Two desks, end to end, stand isolated in the center of the room, with the leader seated facing the door from behind them. File cabinets, tables, and bookcases are overloaded with computer printouts, notebooks, reports, and records of previous experiments. There is a blackboard on the wall, and often photographs or drawings of the group's detectors. The space is crowded with paper, and access is limited.

All eight of the experimental group leaders are Caucasian males; one is Canadian, one is Scottish, and of the six Americans, four are Jewish. The proportion of Jews among leading high energy physicists is often commented upon by Jewish high energy physicists.

Unlike the silent halls downstairs, occupied only by people in transit, the second floor corridors are crossroads, serving as a greeting place. Often physicists stand conversing at one another's office doorways. When major policy decisions are occurring, men stand in small groups, talking and watching.

Aside from people, the corridors are empty except for a few bulletin boards that carry notices of local seminars and lectures, the latest schedules of the allotment of accelerator beamtime, and employee notices. I have never seen anyone besides myself "read" these boards.

Also on the second floor is the library, which houses basic texts, journals, and reference books on subjects directly related to the operation of the lab and its ongoing research. A few general science periodicals are received, as well as a few daily newspapers. A complete record of all reports issued by the lab is also available. All the furnishings (tables, chairs, bookshelves, and carrels) are grey metal. A bulletin board has notices of imminent international, national, and local scientific meetings and lectures, which no one seems to notice. A glass-enclosed booth houses an active photocopy machine, the only one on the entire floor. Behind a counter is a large workspace used by half a dozen library personnel. The area is dense with desks, typewriters, computer terminals, and paper. The head librarian is male; the others are female, one of whom is black; everyone else is Caucasian. The open stock of the library consists of about ten bookcases, each eight feet long and twenty feet high, as well as four large cases for displaying current journals and newspapers. There are four tables with four chairs apiece scattered among the shelves, and tables, chairs, and couches among the periodicals. Physicists are seldom to be found in the library, except at the copying machine or leafing through a magazine in the reading area. I rarely saw more than five or six men in this section of the library, except during conferences held at the laboratory when visitors used the space for studying.

Outside the double doors to the library is a wide corridor, often used for spontaneous discussions between physicists who encounter each other there. On the wall is a bulletin board maintained by the librarians of news clippings about particle physics and other laboratories from both the popular press and general scientific publications such as *Physics Today*. I have seen people glance at these clippings occasionally.

Upstairs on the third floor, one reaches the theoretical physics section, the laboratory directors' offices, and the office of the man who edits and writes most of the articles for *The Beam Line*, the laboratory's monthly internal publication for employees, which describes the physics research at the lab. The editor does an excellent job of describing the work in terms understandable to anyone who has a background in science and is familiar with the lab. *The Beam Line* often reads like an informal historical record of the laboratory. Like many laypeople working at the laboratory, the editor has great reverence for science, scientific discovery, and the laboratory; at the same time, he sees the practitioners as people, not as gods.

Opposite his office is a set of glass doors, which are always open, leading to the offices of the director, associate and deputy directors, and their staff. At least in contrast with the spartan style of the rest of the lab, the director's office is as imposing as the man who occupies it. Like the other directors of major laboratories in particle physics past and present, he is fiercely committed to his lab, with the full force of his personality. The special qualities widely associated with this lab are considered his achievement. Those qualities depend in part on an exceptionally well qualified lay staff, from clerks to technicians. I have heard physicists in the United States, Japan, Europe, and the Soviet Union express envy for the caliber and commitment of the lay staff at SLAC. But there are also more elusive qualities attributed to the staff—often commented upon, but always with a lack of precision that is clearly frustrating to the speakers.

Across the foyer is the office of the associate director, formerly an active experimentalist at the lab. This man assumes much of the responsibility for the routine administration of scientific activities at the lab. For example, he receives all the applications for research associate positions (postdocs) and then coordinates the interview and evaluation process. He is by turns earnest and jocular; he is also clearly devoted to the laboratory. He, the director, and one relatively inactive experimentalist are the only physicists in their sixties at the lab, and only a handful were in their fifties at the time of the study.

In the same suite are the offices of the deputy director, who as a theoretical physicist once made major contributions to the field. While the director represents the laboratory within the physics community and to the federal agencies that fund basic research, the deputy director represents the laboratory to groups outside the scientific community, such as academic organizations and science policy agencies. He is articulate, affable, and experienced at describing research in particle physics in terms understandable to the educated public. These qualities, combined with a commanding personality and strong voice, fit him well for his role as the lab's official elder statesman. During his tenure in this office, he has been active in developing national arms control policy. (It is a commonplace in this community that arms by international agreement must be controlled; on the other hand, particle physicists rarely advocate disarmament.) It would be considered inappropriate for other physicists at the lab to be so active in public affairs. Roles in administration and public policy are clearly to be undertaken after, not during, a major career in research.

From the same large, cluttered office this man also leads the theoretical physics group, which occupies the rest of the third floor. All but one of the nineteen theoretical physicists, including visitors, are male. Their offices usually contain only blackboards and the routine grey metal chairs and desks. From the corridor one hears everywhere the low hum of conversation, the dull scrape of chalk, and occasionally the click of keys from computer terminals. Walking past the offices, one sees the theorists standing in groups of two or three facing their blackboards. One writes and talks while the other, a few steps back, ponders. After a few moments of silence, they trade positions and roles. When working alone, the theorists write seated at their desks; a few work with computers.

Situated in the windowless core of the third floor is the Green Room, which holds about thirty people and is used for small theoretical physical seminars. (Once this room was used for a luncheon honoring one of the founders of SLACWO who was leaving SLAC to become a business executive. The theorists continually peered into the room, their faces registering considerable surprise at the presence of the women.)

One rarely sees experimentalists in the offices of theorists. These people do occasionally discuss physics together, but their meetings typically occur on the second floor. Several theorists told me that an experimentalist would probably feel awkward among the theorists, who have more status. The third floor is very much the domain of the directors, theorists, and their staffs. People based on upper floors freely use the lower floors, but not vice versa.[16]

The physicists eschew any personal decoration or rearrangement of furniture that would differentiate their workspaces. This great visual uniformity, coupled with the clean, functional grey metal and glass decor of the building, creates a strong impression of stoic denial of individualism and great preoccupation with the urgent task at hand. In the use and access to space, there is a sharply defined, nearly military hierarchy between occupational groups and an egalitarian conformity within each group. In the Central Lab one sees little that betrays any interest in individual talent, character, or commitment.[17]

To the right of the grassy square is the Administration and Engineering Building, known at the lab as A & E. It is the most impressive-looking structure on the square, with double staircases leading to the second (main) floor; some structural I-beams give the appearance of a portico of columns. The main entrance, approachable by car, faces away from the square but is even more imposing: a very wide bridge walkway leads to a glass facade. Inside is a large foyer with a great wooden reception desk, leather couches, low tables with magazines, and a relief model of the entire laboratory.

What is significant about this imposing building is the surprisingly insignificant status from the physicists' point of view of what is grouped under its roof. This includes all administrative services, public information, printing, medical care, travel services, and most engineering departments. The "on-site" offices of DOE are also located here. Finally, any groups proposing new research facilities or modifications of the accelerator are housed in A & E until their projects are either rejected or funded. The same suite of offices has been used during this gestational stage by one group after another. All of the above activities are considered support services for the purely scientific work conducted by the people in Central Lab. Senior personnel from A & E occasionally walk to the Central Lab to confer with the director or other physicists there. Most physicists have been in A & E once—to sign the official papers concerning their employment, in the personnel office.

The last building on the square, at the southwest corner, is the Test Lab. The performance of various electronic devices is evaluated here. Some of these, especially the klystron amplifiers, emit a rather annoying continuous sound, audible for about thirty feet from the building. The so-called Health Physics Group is also based here. Its function is to monitor the general levels of radiation at the laboratory, as well as the amount of radiation to which each employee is exposed. This last is accomplished by means of a device called a dosimeter, carried by all those whose work brings them near the accelerator or the Research Yard. The dosimeter is a thick plastic card with the bearer's picture and employee number on it. Wedged inside is a film clip that registers the level of exposure to radiation. Cards are read at different intervals, depending on the amount of time each person is routinely brought into possible exposure. Mine, first issued to me as a tour guide, was read once a year. One must have a dosimeter to pass the guards at the "Sector 30" gate leading to the accelerator and the Research Yard. The guards are there to ensure that all who enter these areas carry a dosimeter. Few administrators or theoretical physicists carry a dosimeter; there are perhaps a score of women who have one.[18]

A road encircles the central square, enclosing the auditorium, cafeteria, Central Lab, Test Lab, and A & E building. Following that road about 180 degrees around the circle from the entrance, one comes to a group of six buildings arranged in a line tangent to the circle. First is the computer complex, which is housed in a facility completed in 1975. SLAC began operation in the early 1960s with an IBM 360/91; in the mid-1970s two 370/168's were added, and in February 1981 an IBM 3081 was installed. At each stage, SLAC's computing power has placed it at the forefront of high energy physics laboratories. This means that, in addition to operating various research devices at the laboratory and collecting data from the diverse experiments, experimentalists at SLAC can also do some "real-time" data analysis. In other words, they can do preliminary analyses of data while the experiment is going on, allowing them to modify the conditions of the experiment to improve the quality of the data.

The Computer Building is massive and looks like a concrete fortress. It was funded during the period of campus political unrest in the early 1970s; in an emergency, all of Stanford University's administrative computing can be conducted from the facility at SLAC, which can be more easily secured than the campus computing center. The people at the Computer Center do not interact often with the physicists, who are primarily interested in having access to on-line data analysis and simply want the lab to have enough computing power to grant them the amount of time they want. The physicists are aware that the hardware and software specialists working on the main computers are very highly qualified people who are themselves experts, working on state-of-the-art equipment. For example, SLAC was the first customer to receive the IBM 3081. IBM keeps several specialists at SLAC to adapt their system to the particular needs of the laboratory. IBM in turn uses the knowledge gained from these special applications to develop new products for their customers with more conventional needs. The computer people at SLAC have their own national and international networks, conferences, and journals to which they submit reports of their research at SLAC.

The other five aligned buildings, taken together, are a highly sophisticated factory for producing many of the electronic and mechanical devices used at the laboratory. The buildings themselves are simple concrete or corrugated-metal structures. The equipment inside is an impressive array of advanced machine tools, along with tools produced at the lab itself. Occasionally, trusted experimentalists are allowed access to these tools. The accelerator itself was produced in these buildings, as well as many magnets and much of the other research equipment. The people who work in this area are engineers and craftspeople who are constantly producing prototypes, translating the physicists' designs into hardware, often developing new techniques in the process. These innovations are reported in yet another set of journals and meetings. It is very important for experimentalists to be on good terms with these engineers and craftspeople, to know whom they can call upon in emergencies. It is this kind of knowledge that gives resident research groups a real advantage over visitor (user) groups and why such groups often employ a liaison physicist at the lab to facilitate their own work. Almost all of the employees in this area (crafts, electronics, fabrication, heavy assembly, and warehouse buildings) are male. There are many female "pinups"; when asked about women working in the area, the men respond that it is not work suited for women because there is little that is routine. The older men, mostly in their fifties and sixties, are Caucasian. Many of the younger employees are black and Spanish-surnamed; SLAC has a rather active technician training program operated in cooperation with a local community college.

Between the computer complex and the manufacturing area is a road blocked by a gate with a guardhouse alongside it—this one with a more than ceremonial function. The guard checks that all who wish to pass have dosimeters, which will measure their radiation exposure in the accelerator and research areas. Beyond this point one is within the areas enclosed by the "radiation fence," which is placed a specified number of feet away from any radiation source. Beyond that gate, extending two miles to the west and twenty-five feet underground, is the linear accelerator. Above ground, all one sees is a corrugated metal structure called the Klystron Gallery, about twenty feet wide, one storey high, and two miles long, with a service road running along both sides; during lunch times the road serves as a popular jogging track.

Walking into the Klystron Gallery one hears the familiar, oppressive sound of the 245 klystron amplifiers, which are placed at forty-foot intervals. The replication of equipment every forty feet for two miles gives the visual impression not of a great vista, as from the central square, but of being in a forty-foot hall with mirrors at either end. All sense of depth is doubled back on itself. A group of BBC filmmakers, producing a show on high energy physics for which I served as laboratory liaison, were able to visually convey the length of the accelerator and the Klystron Gallery only with subtle and difficult lighting arrangements. Behind each of the roughly cylindrical klystrons (about two feet in diameter and six feet high) are large grey metal power modulators that transform low-frequency commercial electric power into direct current and then feed this stored current to the klystrons at regular intervals. The klystrons, which were developed by the Varian brothers in the later 1930s, amplify the weak high-frequency signal to a very high-power burst of high-frequency radio waves (twenty-four megawatts).[19] From the klystron the waves travel through evacuated copper rectangular waveguides, which divide twice to produce four waves from each klystron. These waves descend the twenty-five feet to the accelerator, which is itself a copper pipe about four inches in diameter. The pipe is actually a series of cylinders about four inches across and one inch thick separated by thin disks with small holes in the center. (It be can be visualized as a two-mile string of pet food cans with one-inch holes in the bottom, brazed together to form a pipe.)[20]

The "injector gun" at the beginning of the accelerator in the west produces and aims electrons into the copper accelerator. It also "bunches" the electrons and positrons, each bunch according to specified parameters. The radio waves (2,856 MHz) entering the accelerator from the klystrons at ten-foot intervals "push" the electron bunches for ten feet until the next wave enters the accelerator. The water-cooled accelerator itself must be perfectly aligned so that the "beam" (the collection of electron bunches) can remain straight. To accomplish this, the accelerator is mounted on an aluminum pipe, two feet in diameter, containing Fresnel lenses. These lenses are used to align the accelerator; the beam itself is focused by magnets every 333 feet.[21]

The entire apparatus stands on the floor of a tunnel eleven feet high and ten feet wide, with thick, steel-reinforced concrete walls, which is separated from the surface level by twenty-five feet of earth—all of which is for radiation protection. No one is allowed into the tunnel while the accelerator is operating; a system of safeguards would cause the accelerator to turn off if anyone were to try to open one of the doors between the Klystron Gallery and the accelerator tunnel. The tunnel was also specially constructed to withstand earthquake movements.[22] For stability, the two-mile tunnel floor has no expansion or contraction joints. New construction techniques were developed for this project: eighty-foot-long sections were poured, mixed with ice to delay curing while the next section was poured. Especially developed techniques for evaluating building sites near earthquakes were necessary for government approval of the 480-acre Stanford site. Construction of the accelerator was authorized by Congress on September 15, 1960, with an estimated cost of 114 million dollars; the first beam was accelerated on May 21, 1966.[23]

At the eastern end of the two-mile accelerator, near the guardhouse where radiation badges are checked, is a large fan-shaped mound of earth. In this region, one thousand feet long, the discrete bunches of accelerated electrons are steered toward one of six different experimental areas by means of large pulsing magnets. This area, called the Beam Switchyard, has extra-thick concrete walls and earth covering because of the increased radiation. The switchyard, under fluorescent lights, looks like the interchange in a subway system with large rails upon which the magnets are moved. It is very cool and quiet: a nice place to sit on a hot day.

The beam switching, along with accelerator operations, is managed from the Main Control Center by means of PDP-9 and SDS-925 computers and some auxiliary PDP-8's. The people at the Main Control Center developed new devices, such as "touch panels," to facilitate their operations.[24] The Center, a simple metal building, looks on the inside like the cockpits of ten jets combined. It is an entertaining array of flashing colored lights, CRT displays, and buttons, all framed by cartoons and posters above the control boards and great tangles of computer cables on the floor. In addition to the mechanical processes they oversee, these operations are sometimes the focus of intense pressure from physicists. When their share of the beam is not delivered exactly as promised, the experimentalists complain vociferously. On the other hand, if their research equipment malfunctions for a while so that they do not use all of the beam allotted to their experiment, the physicists may try to cajole the operators into a share of the beam later on. The director of the Main Control Center has a sense of humor and a strong sense of the integrity of its activity. He trains the operators on the equipment and has promoted some women from clerical positions to the higher-paying position of accelerator operator. It is one of the few occupations at the laboratory held by both women and men in significant numbers.

Beyond the Beam Switchyard, the road descends into a large area that has been excavated twenty-five feet—down to accelerator level. This fan-shaped area (about fifteen hundred feet in radius, with a ninety degree arc) is the Research Yard. The whole area is paved and surrounded by the unexcavated remains of the hill that once covered it. The impression is one of standing in a flat-bottomed, high-rimmed crater; the only view is of the sky. The yard was designed so that the earth enclosing it would absorb any ambient radiation, but the beams as they enter from the switchyard are relatively unshielded. Some theorists will not go into the Research Yard while the accelerator is running and experiments are in progress; they consider the radiation levels there to be hazardous. I have never heard experimentalists discuss this. According to the Health Physics Officer at SLAC, which monitors radiation exposure on site, the amount of radiation at SLAC is less than that at many industrial plants in the region. The officials at the Stanford University Fire Department substation at SLAC consider the threat of fire to be greater than any radiation problems.

The fear of contamination from radiation emitted by the accelerator and experiments at SLAC has been a constant source of laboratory-community tension.[25] The "radiation fence" is five hundred feet from the research and accelerator areas, completely enclosing them. To SLAC, this fence represents the solution; to the local residents, it represents the problem. The federal government established that maximum radiation exposure near such facilities should be no more than double the amount people are exposed to "normally" (which includes both natural sources, such as cosmic rays, and synthetic sources, such as X-rays). SLAC placed the fence at a distance such that a person standing there for a full year would receive one-half this "doubling dose."[26] The safe level of exposure to radiation is much debated.

Many cultures identify substances or activities as polluting if they cannot be easily classified into some bounded system.[27] Often, actions and things that are forbidden are also sacred: in this case, radiation sources are seen as both necessary for basic knowledge of nature and carcinogenic.

> In [observing the rules of cleanliness] we are not governed by anxiety to escape disease, but are positively re-ordering our environment, making it conform to an idea . . . The ideal order of society is guarded by dangers which threaten transgressors. These danger-beliefs are as much threats which one man uses to coerce another as dangers which he himself fears to incur by his own lapses from righteousness. They are a strong language of mutual exhortation. At this level the laws of nature are dragged in to sanction the moral code . . . The whole universe is harnessed to man's attempts to force one another into good citizenship . . . I believe that some pollutions are used as analogies for expressing a general view of the social order . . . Ideas about separating,

> purifying, demarcating and punishing transgressions have as their main function to impose system on an inherently untidy experience . . . Reflection on [pollution] involves reflection on the relation of order to disorder, being to non-being, form to formlessness, life to death.[28]

The SLAC radiation fence is a tangible border, but it does not confine the radiation, although the potency does diminish with the distance from the source. It is a symbol of the lab's restraint, its responsibility in its dealings with the community; it is also a symbol of the lab's very great and dangerous power. Conversely, the designer-green power lines carrying energy into SLAC are a reminder of SLAC's dependence on the society which supports it.

There are no fences around the major high energy physics laboratory in Japan, KEK (*K*o-*E*nerugie butsurigaku *K*enkyusho). It is part of a new "science city," Tsukuba, which includes many scientific research labs, a university, and housing for the staff of all these institutions and their families.[29] Many Japanese people say that Tsukuba Science City is "not really Japan." I think that Tsukuba represents a distinctively Japanese experiment in the organization of basic scientific and engineering research. There is nothing like it anywhere in the world. It is disturbing to many people in Japan, I believe, because it breaks with some Japanese traditions in science and engineering, in work and in education. It is also a very new town in a largely agricultural district; most Japanese seem nowadays to prefer the huge, established cities, especially Tokyo. Certainly there are many frustrations and difficulties here for both the Japanese newcomers and the visiting foreigners; there are also immeasurable opportunities. Americans would call it a "frontier town."

Part of Tsukuba and part of KEK seemed quite familiar to me after my sojourns at SLAC and Fermilab, but the familiar parts were strangely juxtaposed with other features that seemed very Japanese to me. This reminds me of visiting Nariaiji, a temple near Amanohashidate on the western side of Honshu. The temple was isolated and nicely situated on a mountain top. It took me a long time to get there. Everything was as I had fantasized the "ideal" Japanese temple: multicolored streamers marking the long staircase, remarkable wooden architecture, incense burning in a great iron pot, dimly lit golden images, a young monk selling for a few yen calligraphed papers imploring one to "destroy the self," and the whole front of the temple wrapped in a banner of white chrysanthemums on a purple background. On the porch of the temple—to the left of the incense burner and directly opposite the young monk—was a bright red Coca-Cola vending machine. Behind the temple, near the living quarters, a monk was filling the tank of a large black limousine from the temple gas pump. It took me some time to realize that Tsukuba and KEK were much less familiar than they seemed at first glance, less American and very Japanese.

From Mount Tsukuba, I could see the university and the surrounding farming communities. A few of the towns and villages have important religious shrines, and some, such as Mashiko and Kasama, are centers of traditional Japanese pottery making. The laboratory site was visible, too. The layout of the laboratory on its site was quite familiar, yet the allusion escaped me until the day I bought a piece of pottery in Kasama. Returning to the visitors' trailer at KEK where I was staying, I unwrapped the box containing my purchase. Wrapping things is given very special attention in Japan. For each wrapping material, there appears to be a canonical method. For things wrapped in paper, a rectangular piece is always used; the container to be wrapped is aligned with an imaginary diagonal line on the paper. I have seen very young sales clerks make two or three tries at wrapping a box until they have found the "right" diagonal reference line. The finished product is a wrapping with folds only along the edges of the container, and only one exposed paper edge which cuts a diagonal across one surface. The wrapping is held securely closed by one sticker, usually decoratively embossed, placed upon this final single diagonal. Whenever I opened one of these carefully wrapped parcels, I studied the creases on the paper to see if I could duplicate the process. The creases formed a series of diagonals and triangles (where pleats had been made at the corners).

The creases in the paper wrapping the pottery from Kasama strongly recalled the site plan of KEK, with the two main office and laboratory buildings and their shared reflecting pool aligned on that initial diagonal. The remaining rectangles and triangles are filled with laboratory buildings, parking lots, and open land.

There are many evergreen trees and thick, shiny high grasses around the lab in the spring. As I walked from the guest trailer to the main building, it was often difficult to find the path in the luxuriant, sweet-smelling grass. Some of the people at the lab had cleared a space in the field for a volleyball court, which was as busy at noon as the nets at SLAC. At both KEK and Fermilab, near Chicago, the site is flat and the mounds of earth covering the accelerator and switchyards provide the main diversity in the immediate landscape, aside from the multistoreyed laboratory buildings.

KEK is an active, busy place, and the buildings are light and spacious. This was in sharp contrast to the university physics departments I had visited in Japan. University offices were small and neat, often with elevated tatami mats and a place near the office door to leave one's shoes; but the halls were crammed with filing cabinets and research equipment; they were also often unkempt. This jarred one of my cultural stereotypes. I learned that each university department's budget includes money for keeping the shared space clean. A department can decide to use those funds in other ways, keeping only the most rudimentary janitorial services. In Japan, as in England and Europe, spartan, barely maintained schools can symbolize a studied disregard for material life and a commitment to intellectual pursuits.[30] By the time I saw Tohoku, the new university at Sendai, and the new laboratories at Tsukuba, I was able to be startled by their very clean and very modern large, light spaces.[31] They seemed very stark and foreign to me.

Inside the main laboratory building are offices and the library. I used the library as my office: a place to write field notes, describe interviews, and study the documents produced by the lab, as I had done at SLAC and Fermilab and many research centers in Japan. As always, the regular library staff were women, and most of the library users were young men. At Nagoya University, when I first arrived, a senior physicist pointed out to me a very good collection on the history and philosophy of physics in the department library—the result of his own strong interest in the subject. That evening at dinner with three faculty members and four graduate students and postdocs, all men, one cautiously asked if I had found the library useful. I said yes, adding that I had found a book there which I had been wanting very much to read. It was a collection of articles written by Japanese historians and sociologists of science as well as physicists, on the subject of science (primarily physics) in Japan;

the collection had been gathered, translated, and published to coincide with a meeting of an international history conference in Tokyo.[32] As it was passed around, everyone at the table looked at the book carefully, and each seemed very surprised, though I could not say why. Finally one of the postdocs asked me, "How did such a beautiful woman come to study physics?"

Neither in my home culture nor in Japan could this question be taken at face value. I had learned already that my red hair and green eyes remind many Japanese of a monster called Tengu from a folk tale often told to children. When I went marketing, children would sometimes fall off their bicycles in surprise if they saw me unexpectedly; I could hear them murmuring "Tengu!" as they lay crumpled in a heap, staring up at me. I was well aware that my presence in Japan as a researcher, my subject area, and my being a woman surprised the physicists. In addition, I was in Japan with my (now former) husband. For a woman to continue working after she is engaged to be married was still relatively rare in Japan at the time. Finally, in Japan when men who work together gather to eat dinner, their wives do not join them. When women are present with men on formal occasions, they do not speak often; it is men who have the responsibility in mixed company for establishing and maintaining social conversations. For all these reasons, my presence in the physics library and at dinner, my conversation and my appearance, would all be quite foreign. When I arrived at KEK I discovered that my presence at the lab and my researches surprised no one. I learned that many of the physicists at KEK have studied or even worked abroad for extended periods; they were more familiar with the existence and behavior of American women researchers.

In university departments, a senior professor's office is crowded. It usually includes his desk, bookcases, his secretary's desk, a table for discussions, and a small couch and side table, and it serves as a gathering place for the whole research group. My first meeting with a senior university professor (arranged in advance by telephone) began with me being seated at the small couch. The rest of the group would join us; then the secretary brought tea on a tray and served a cup to each. I was quite embarrassed to discover, well into my first such interview, that I was the only one drinking the tea; I had not realized the tea was to be regarded only as a social gesture. Once the men gathered, I would begin to ask questions and they would ask me how things were organized at American universities and laboratories. Such talks usually lasted about one-and-a-half hours.

At KEK I had had difficulty arranging a meeting with one research group leader because the whole group was in the process of moving to another set of offices. When I came across him pointing here and running there, directing the move, I was sure he would not want to stop for our scheduled talk, but I was mistaken. He announced to all within earshot that he was going to be interviewed by this young lady, and that he would be in such-and-such an office across the hall with me if he were needed. He seemed to enjoy all the attention this brought him. We sat in the makeshift office and began talking immediately. At KEK, Tokyo University, and Tohoku University at Sendai, I was able to have prolonged conversations with individuals, but meeting with groups of people was difficult to arrange. The opposite was true at other Japanese universities and institutes.

The accelerator at KEK is a proton synchrotron. Proton accelerators, as the name suggests, accelerate protons, which, along with the neutron, form the nucleus of the atom. Protons are thought to contain many constituent particles; when the accelerated protons collide with targets, these particles form part of the resulting debris. Usually protons are accelerated in a circle, as at KEK and Fermilab, increasing their energy with each circumnavigation; in a synchrotron, their path is held at the proper radius by magnets, which increase the power of their magnetic fields as the energy of the protons increases. Protons must be preaccelerated in smaller, less powerful accelerators before being "injected" into the synchrotron, with its rather large radius. (At Fermilab, the 400 GeV proton accelerator is two kilometers in diameter; the diameter of the KEK machine is 108 meters.)

As particles accelerate (whether they are protons or electrons), they radiate energy in the form of photons at a rate proportional to their velocity, which is both increasing and changing direction in the curved path of the accelerator. Protons have 1,800 times the mass of electrons; if an electron is accelerated to the same energy, its velocity is that much greater than the proton's, producing more radiation. Because of difficulties in controlling the radiation, electrons are now usually accelerated in the more expensive linear accelerators (or "linacs"), such as SLAC's, in order to reduce the radiation loss at high energies.

The KEK synchrotron is composed of four accelerators, a 750 keV Cockcroft-Walton preinjector, a 20 MeV injector linac, a 400 MeV fast-cycling booster, and the main 12 GeV synchrotron.[33] The Cockcroft-Walton was brought to KEK from Tokyo University. The linac, booster, and main ring are new. The components for the accelerator were constructed by private industries. Because Mitsubishi had special capabilities in copper-plating technique they constructed the linac. The magnets were built by Hitachi and the ion vacuum pump by Nikon Varian. Construction on KEK and the accelerator began in April 1971; an 8 GeV beam was achieved in March 1976, shortly before my arrival.

In 1965 the accelerator was designed to reach 40 GeV, but in 1968 the budget was cut to one-quarter of its projected thirty million yen funding. The accelerator has been designed to accommodate extensions to higher energies, should the physicists be able to get more funding. In order to gain that funding the Japanese physicists needed to become more familiar with the corridors of power in Tokyo. An American ethnographer studying geisha life told me that she had served at dinner meetings of businessmen, politicians, and basic-research scientists. She said that the scientists had difficulty participating in the informal banter and subtle negotiations which are the substance of such meetings.[34]

In her excellent account of the development of KEK from 1959 to 1971, Lillian Hoddeson argues that "common structural features contributed to Fermilab's and KEK's development in the period 1959–1964 on strikingly parallel tracks . . . But the histories . . . diverg[e] notably . . . in the period 1965–1970."[35] This divergence was due, in part, to the differences between American and Japanese science funding politics and the organization of the physics community in each country. It is to the difference in machines that I now turn.

CHAPTER · 2

Inventing Machines That Discover Nature: Detectors at SLAC and KEK

Most people, when they think of tools for scientific research, think of telescopes and microscopes and X-ray machines; they assume that research equipment clarifies what is remote, small, or otherwise inaccessible to sight, much as eyeglasses help some of us to see the world more clearly.[1] Even accelerators have sometimes been likened to big microscopes. To understand the role of research equipment in high energy physics, one must have some familiarity with the stages of the research process.

It is an ancient idea that the universe is composed of indestructible fundamental elements and, nevertheless, remains infinitely complex and perpetually in flux. Classical Chinese, Hindu, and Greek texts disclose debates about the nature of these raw materials of the universe. In the late nineteenth century the contemporary model of the basic elements (called atoms, after the Greek word meaning indivisible) began to take form. J. J. Thomson proposed that atoms were composed of clusters of particles with negative electrical charge; the particles came to be called electrons (from a Greek word meaning amber, which retains an electrical charge). In 1913 Ernest Rutherford concluded that the bulk of the atom's mass was concentrated in a positively charged core, the nucleus, which is surrounded by orbiting electrons. He identified the positively charged constituents of the nucleus, named protons (Greek for *first* or *primordial*), in 1921; James Chadwick established in 1932 that the nucleus also contains particles without charge, or neutrons.

Even this tripartite model of the atom has not remained stable.

Atoms routinely emit and absorb electromagnetic radiation in the form of photons. A decaying neutron emits a neutrino as well as an electron and a proton. Interacting nuclei exchange pi-mesons. Further discoveries in particle physics have increased the number of "elementary" particles to well over one hundred. Various models have been devised to describe the relationships among these many particles. The classification system devised by Yuval Ne'eman and Murray Gell-Mann in 1961 (named the Eight-Fold Way) organizes the particles according to mass, charge, decay products, spin, and so forth. This system is analogous to the periodic table of the chemical elements developed by Mendeleev in the nineteenth century.

In addition to identifying the basic materials, any theory of fundamental immutable elements must also account for the appearance of movement that the universe presents to us. Historically, in physics this question has been posed in the following way: How do bodies act upon one another across space? Modern physics has identified four forces to account for "action at a distance." Two of these, gravity and electromagnetism, are familiar to classical physics. Two others, the strong and weak nuclear forces, are not.

From the work of Newton, Laplace, and Einstein the force of gravity is rather well understood; it operates from the cosmic to the terrestrial scale. Electromagnetism operates on the astronomical, planetary, terrestrial, and atomic scale. The strong nuclear force was first suggested in 1933 to explain how the positively charged proton and the neutron were bonded together in a stable nucleus on a scale where gravity is weak and electrostatic repulsion ought to force them apart. Physicists are able to manipulate this strong force, but it is not completely understood. The weak nuclear force was proposed in 1933 by Fermi to account for the natural radioactive decay of nuclei. This theory correctly predicted the existence of the neutrino.

Developments in particle theory forced physicists to reevaluate fundamental tenets of classical physics. From the sixteenth century to 1900 physics supported and refined a mechanistic view of nature. Classical physics, furthermore, had as its foundation an experimental method that stressed isolating cause from effect and determining which specific cause produces which particular effect. In 1926 Heisenberg and Schrödinger rejected the notion of strict causality in atomic processes. They concurred that at any given time it is impossible to determine the exact state of a system. If we cannot determine the exact state of a system at two separate points in time we cannot strictly determine specific causes and particular effects. We can only project *statistically* what effect might succeed what cause. The uncertainty principle was a serious challenge to the mechanistic model of classical Newtonian physics, and replaced it with a statistical model of the universe.

Initially in particle physics electromagnetism, gravity, and the strong and weak forces were seen as incommensurable. Then, electromagnetism and the weak force were linked by Sheldon Glashow, Abdus Salam, and Steven Weinberg in a theory of the electroweak force; this theory correctly predicted two new particles, the *W* and *Z*, which were found in 1983. Eventually gauge field theories of the strong force and the electroweak force were combined into a Grand Unification Theory, or GUT. Theoretical efforts are now under way to incorporate gravity with a GUT, and these are called "superstring theories." Physicists are also trying to incorporate superstring theories and supersymmetry into a "Super GUT." The proposed research device known as the Superconducting Super Collider (SSC) is justified as necessary to investigate these new theories.

The SSC is an accelerator, which is a device that impels particles to very high energies and then directs them at "targets" composed of other particles. The collisions of the accelerated particles with the target particles generate energy; most of this energy takes the form of new particles, although some energy is released in the form of radiation. Targets are closely surrounded by diverse devices that record traces of the new particles. The target plus the recording device and the computing system that analyzes the records is called a "detector."

At any laboratory there are many detectors near the accelerator, receiving accelerated particles from it. Since the particles are clustered into discrete pulses, they can be directed—by magnets in the switchyard that bend their paths—to any of the detectors. One detector may be able to accommodate sixty pulses per second; another, ten; yet another, thirty. The full load of the accelerator beam (which may be as much as three hundred and sixty pulses per second) is distributed in this way to the several experiments being conducted concurrently, each using one detector.

The accelerator belongs to the laboratory as a whole, but each of the resident groups of experimentalists conceives, constructs, maintains, and develops its own detector. Detectors are distinctive and serve as the "signature" of the group. These machines are at the heart of the research activity of particle physicists. Detectors—their conception and development, their maintenance, their performance during the precious allotment of beamtime for an experiment—are the stuff of frustration, hope, heartbreak, and triumph for research groups. Discovering a new way to detect and record the traces of elementary particles can bring great honor and influence. A detector that ran perfectly at all times would be considered either obsolete or not daring enough in conception.

The place of detectors in high energy physics contrasts sharply with the role of research equipment in many other fields. In much biological research, for example, laboratory machines are not built or even designed by the scientists using them; they are mass-produced and advertised in catalogs. Their design incorporates theories and laboratory practices so widely accepted that their validity has not been questioned for many years, perhaps decades. Bruno Latour and Steve Woolgar in their study of an endocrinology laboratory vividly describe how such tools represent "reified theory" and are regarded by scientists as useful, but routine and uninteresting.[2] The endocrinologists' research reports never describe the workings of the "black boxes" they use.

High energy physics detectors are not black boxes with unquestioned assumptions hard-programmed into them. In high energy physics inventing machines is part of discovering nature. In this chapter I describe four styles in detector design, represented by three detectors at SLAC and one at KEK. These four styles represent four different strategies for making discoveries: four different strategies for spending money, organizing a research group, and becoming a success in science. Many physicists have claimed that big machines in science determine the organization of scientific research, even that they determine the research questions. By looking closely at how specific machines shape scientific practices and at how scientists regard the machines they are constantly building and rebuilding, we can challenge the claim of mechanistic determination.

Most high energy physics detectors share certain characteristics. A closed container houses a highly sensitive medium. Pulses from the accelerated beam enter the container and collide with particles of the medium; new particles, generated by the collisions, are propelled through the medium, creating disturbances that are meticulously tracked and measured. The record of the pattern of disturbances is then analyzed for clues to the properties of the particles that caused the disturbance. Significant differences between detectors of the same type depend upon how sensitive the initial environment is, how effectively that environment can be controlled, how finely the disturbance of the medium can be differentiated, and how precisely that disturbance can be calibrated. Physicists also want to know how frequently the process can be repeated so as to accumulate as many events as possible during each experiment; this will enhance the validity of their conclusions and increase the probability that a rare event will be "seen."

Other aspects of the ambient environment, besides the particles generated by the controlled collision, also disturb the medium and trigger the measuring and recording equipment. These undesired disturbances are "noise." A fine detector should enable physicists to identify and measure the noise generated in any specific experiment. A highly sensitive medium, capable of finely differentiated disturbance, associated with a system that can measure those minute fluctuations, adds up to a delicate piece of machinery, quite vulnerable to high levels of noise, some of which will be unpredictable. Experimentalists strive to maximize sensitivity and speed while minimizing noise, especially unpredictable noise. In the high energy physicists' view a mass-produced detector would necessarily have an unacceptably large margin of conservatism built into it. Once one of their own detectors begins to work with great regularity and predictability, it becomes a candidate for manufacture and distribution to scientists in other fields; it is obsolete for high energy physics. A detector can be discarded (or never even built) for another reason: cost. The price of achieving the sensitivity, calibrating the noise, or "reading" the data can become prohibitive.

Given these common features and problems, detectors are rated by the extent to which they maximize one of the component variables: sensitivity in identifying the presence of particles, speed of data collection, capacity for distinguishing noise, and efficiency of data analysis. Bubble chambers provide the most elaborate data on particle behavior; but they collect data much more slowly than other detectors, and since the data are recorded photographically rather than directly in computers, analysis involves human scanners and is therefore lengthy and costly. At the other extreme are "counters," which merely signal that some particle has passed through a sensitive grid of wires at a specific point; this sort of information can be recorded immediately in computers and analysis can begin while the experiment is still running so that the experimentalists can judge the quality of data they are gathering and alter the experiment accordingly; this is "on-line" or "real-time" data analysis. Experimentalists always search for new ways to collect complex data quickly so that information can be recorded directly into computers for on-line data analysis.

Ancestral Machines

There have been nine detectors in the SLAC research yard, fed from the SLAC accelerator: three bubble chambers, a spark chamber, a streamer chamber, a large aperture solenoid spectrometer, a group of three spectrometers, and two detectors associated with a colliding beam facility. The bubble, spark, and streamer chambers represent refinements of decades-old innovations. The others are examples of more recent developments. Before turning to a discussion of the four styles of detector design I came across during my research in the 1970s, I will briefly describe their forerunners.

Bubble chambers were developed by Donald Glaser at the University of Michigan in the early 1950s; in 1960 they brought him the Nobel Prize.[3] He is quoted as saying that he decided to persevere in his studies on bubble chambers because of a conversation over a few beers with fellow physicists:

> After several pitchers of beer we began to wax philosophical about physics. One of the boys, looking dreamily into the pitcher of beer before him, saw the usual streamers of bubbles and remarked, "You can see tracks in nearly everything." Just for fun I actually exposed some beer to gamma rays the next day in the laboratory. Nothing happened.[4]

Bubble chambers usually contain hydrogen, which has a comparatively simple structure. Liquid hydrogen in a very smooth vessel is put under increasing pressure; under certain conditions, it can be "superheated" beyond its boiling point without boiling. In this state it is highly sensitive to any fluctuations in temperature. Then, a bunch of particles from the accelerator is allowed to enter the vessel. Each collision between an accelerated particle and a particle of the superheated liquid generates heat, causing the liquid to begin to boil at the point of collision; the pressure is lowered just enough to allow an initial bubble to form. The energy created by each collision re-forms into new, short-lived particles, which scatter away from the original collision; the heat generated by their movement causes the liquid to bubble along their trajectories. The entire vessel is surrounded with strong magnets, which are turned on just as the initial interaction occurs. This causes any new particle with an electrical charge to be attracted toward one or the other of the magnetic fields; its bubbling track bends accordingly. The angle to which that particle's trajectory is bent by the known force of the magnetic field is a measure of the particle's momentum. If a particle disintegrates or decays into other particles while in the medium, the trajectory will split. The whole array of bubbling tracks in the bubble chamber is photographed from three directions so that the trajectories can be reconstructed in three dimensions. Then the pressure must be further reduced to allow the bubbles to recondense back into the liquid; if the whole liquid in its superheated state were to start to boil, the vessel might well explode. When the liquid regains its stability, the whole process is repeated. Bubble chambers vary in how often this process can be repeated and in the quality of the pictures that are produced. (One bubble chamber at SLAC, the eighty-two-inch chamber, took 24.3 million pictures over a five-and-a-half year period, pulsing twice a second during its experimental runs.)[5] The pictures are then scanned to measure and classify all the tracks, and this information is recorded in computers and analyzed for "significant events"—tracks with particular configurations of interest to the current experiment. Two advantages of bubble chambers are that resolution of the particle track is very good and that the liquid can be changed to provide alternative target particles.

Spark chambers and streamer chambers are similar to bubble chambers in that a sensitive medium is used to signal the presence of a particle and its path.[6] Both spark and streamer chambers rely on a stack of two-dimensional grids of fine wires aligned in the sensitive medium; when the grids "fire" in sequence, the inference is that one particle and its decay products, passing through the grids, caused the series of signals. Spark chambers can operate much faster than bubble chambers, and they can be set up so that photographs will be taken only when significant events occur; this reduces the time and cost of data analysis. Streamer chambers represent an attempt to combine the advantages of bubble and spark chambers—selective triggering on the one hand, and fine resolution of tracks on the other—by turning a series of isolated sparks into a stream of dots, as in a bubble chamber picture. The result is a set of discontinuous signals as in a spark chamber, not a continuous track, as in a bubble chamber: the grids of fine wires can never be placed close enough together to replicate the density of signals possible in liquids.

SLAC's spark, streamer, and bubble chambers are fine examples of their kinds. The eighty-two-inch bubble chamber was built, in fact, at the Lawrence Radiation Laboratory in Berkeley in 1959, as a seventy-two-inch chamber. In 1967 it was brought to SLAC, with most of its operating crew, to take advantage of the special characteristics of the accelerator beam there. The chamber was dismantled, modified, and rebuilt as an eighty-two-inch chamber. It was dismantled finally in 1973 because the cost of analyzing its photographic data, relative to other detectors, had become prohibitive. The director of the laboratory also argued that, although the eighty-two-inch chamber was a detector that still produced very fine physics, it was no longer sufficiently distinctive. His strategy was to maximize the laboratory's access to funding by concentrating on research equipment and processes that could not be duplicated at other laboratories. I attended the party held in the fall of 1973 in honor of the chamber, just before it was finally dismantled.[7] The twenty-five-foot-high structure housing the chamber was decorated with streamers from reels of film which had been used in recording its data. Among the people present were those who had been involved in building, maintaining, and modifying the chamber and many of the physicists who had designed it and used it for experiments. Seventeen new particles and resonances had been discovered in the course of its long and honorable career.

The physicists and operators of the eighty-two-inch chamber were disturbed and saddened by its impending demise. We stood in the metal shed, gathered around the machine, drinking beer (of course) and listening to tales of the glories and quirks of the fourteen-year-old machine. The technicians at the chamber had been among the first at the lab to befriend me and explain their detector to me in detail; I had long been impressed with their ingenuity in simultaneously modifying the detector and keeping it working so well. It looked like a Rube Goldberg contraption and made a lot of noise while it was operating. The pressurization system was built partly with Navy surplus materials from submarines, partly with the most sophisticated special-purpose fittings available. The cameras used to record the tracks had had to be rebuilt at SLAC because the specifications could not be met by any supplier. The group members were proud that they had expended money and time selectively, only when needed, and used experience and ingenuity to find workable shortcuts.

I had especially enjoyed standing above the chamber while it was operating, peering inside its windows and watching the tracks form and re-form as the whole structure shook with each pulse of the pressurization system. The structure felt rickety and looked worse, but the chamber itself, encased in all its supporting equipment, was gleaming, smooth, and elegant. The events in the chamber gave me the impression I was actually seeing the fundamental constituents of nature at work, as if through a gigantic microscope. It was not easy to remember that I was looking at signals produced by a machine designed to react in a stylized way to the debris from a collision occurring under highly controlled and artificial conditions. I was not looking through a window on the world of subatomic particles; I was doing something more like reading dinosaur tracks. It was the "real-time" activity in the chamber that created the impression of seeing.

The group had already begun work on two new bubble chambers. They had "married" a forty-inch bubble chamber to a spark chamber and achieved a pulse rate of six pulses per second. By using the spark chambers to pick out interesting events, they were on average taking only one picture in thirty pulses; the rapid pulsing meant that many more interesting events were observed, and the bubble chamber provided rich data on each. Later the hybrid forty-inch chamber achieved higher pulse rates and was "married" to other devices used to refine its selection of interesting events to photograph. Its pressurization system had been cannibalized from the eighty-two-inch chamber.[8] Groups can cannibalize their own past, but they cannot take over equipment from other groups, no matter how useless the materials may have become to their owners. The same holds for computer software. I found this to be true in Japan, at Fermilab, CERN, and SLAC.

To give a sense of how sensitive a detector is and how a group is preoccupied with designing, maintaining, and modifying their detector in addition to operating it for experiments, I will quote at length from a report written by a member of the Bubble Chamber Group on the hybrid forty-inch chamber:

> A number of changes were made to shorten the recondensation time and extend the pulse rate. Pressures were raised to cause more rapid condensation. Heat exchangers were positioned to promote convection and prevent bubbles from collecting in pockets. Cracks and crevices, the most copious producers of bubbles, were eliminated wherever possible. There are no longer any bolted joints inside the chamber. Everything has been welded. The main glass window seal, a major source of bubbles, now has a special cast and machined indium sealing surface which is far superior to the conventional indium wire seal.
>
> A second problem is to produce accurate and uniform expansions with the hydraulic actuator. The piston must move with a precision of a few thousandths of an inch in space with an accuracy of one ten thousandth of a second in time. The hydraulic system must produce this motion against a three-quarter ton force ten times per second.
>
> When everything is in good working order, the expansion system, originally developed by the operations group for the 82-inch bubble chamber, works admirably well at 10 pps [pulses per second]. There are some difficulties, though. At 10 pps mechanical vibrations do not die out between pulses but build up to such a level that parts break from fatigue. In spite of a continuous program of strengthening, a large number of bolts were broken during the experiment. Repairs were often made to the machine while pulsing in order to keep it going. Ten expansions per second require 30 gallons of hydraulic oil per minute under 3000 pounds per square inch of pressure. During the early part of the experiment there were numerous failures of the hydraulic plumbing. With this quantity of high-pressure oil flowing, the leaks were often spectacular.
>
> Once the chamber began to run regularly at 10 pps, the pictures began to show that some tracks were disappearing in parts of the chamber and reappearing in other parts. At first it was thought that this was due to incorrect temperatures in some parts of the chamber which prevented the bubbles from growing. This had to be discounted when it was found that the trackless areas varied from pulse to pulse. Furthermore the problem went away if the chamber was pulsed slowly. The difficulty was finally uncovered

> in the hydraulic system. At high pulse rates, oil in the exhaust pipe would resonate like air in an organ pipe. A pocket of gas would collect inside the actuator. Each time the chamber pulsed, this gas pocket collapsed causing a vibration that was transmitted to the main piston inside the chamber. The pressure variations in the hydrogen from this piston vibration caused the track bubbles not to form in some parts of the chamber. Once the oil pipe was pressurized to prevent the gas pocket, the vibration disappeared and tracks were visible throughout the chamber.
>
> The entire experiment was done with one Scotchlite reflector. The chamber stayed clean for the whole experiment and the optical quality of the pictures was always good. This experiment would have been impractical in any chamber which became dirty from pulsing.[9]

The other new machine the group was beginning to work on was the fifteen-inch Rapid Cycling Bubble Chamber (known as "the RCBC"). It had an electromagnetic rather than a hydraulic drive system and would ultimately be able to pulse up to sixty times per second. Its powerful magnet, which was nevertheless quite small because it was superconducting, was designed and built by the Cryogenics Research Group at the lab.[10] These newer bubble chambers are smaller than the eighty-two-inch and are contained in more compact supporting structures.

Machine Styles and Strategies for Making Discoveries

The ESA detectors. As one travels beyond the stately central plaza at SLAC, beyond the uniform laboratory offices, beyond the orderly two-mile procession of klystrons accelerating properly aligned electrons, beyond the subterranean, systematic beam switchyard, one descends by road into the huge, paved Research Yard, excavated twenty-five feet to accelerator level. The Yard resembles nothing so much as a big, messy manufacturing business. At one end is a monolithic, windowless concrete structure, seven storeys high. Several corrugated metal one- and two-storey buildings are scattered around the Yard. The rest is a haphazard jumble of massive concrete blocks, large cable spools, stacked lumber, very big cranes, fork lifts, some toilet sheds, and clusters of automobiles. The people working here look a little dirty; some wear safety helmets. Over the years I have discovered that most non-scientists think of labs as extremely clean, meticulously tidy places where people in immaculate white coats do their work with minute, precise movements, and that scientists work alone, in silence. High energy physics laboratories are not like that.

Walking around this terrain, one is confronted by obstacles in almost every direction—raised wooden burrows housing the electrical cables, thick as a fist, that connect the detectors to power sources and to computers. One crosses these by means of little ladders, much like stiles for crossing country fences.

At the far end of the Research Yard is "End Station A," a massive seven-storey concrete structure with thick walls that move on rails like barn doors. The heavy concrete protects against the radiation produced inside. As with the accelerator and the switchyard, no one is allowed in this building during an experiment; the research devices are controlled remotely during a typical "run" of several weeks. Radiation is a problem at End Station A because the spectrometers there do not retain all the new particles generated from collisions; bubble chambers do retain all the generated particles, so the radiation they produce is contained.

Experiments here can make use of the maximum energy of the beam and up to 320 of the beam's 360 pulses per second. When the incoming accelerated electrons collide with target protons, the electron is said to "scatter" and the proton to "recoil" ("elastic electron scattering"); in some interactions new particles are also generated ("inelastic electron scattering"). In some experiments the electron beam is directed at a preliminary target, generating photons which are used as a "secondary" beam to collide with the usual target protons. In this case (photon-proton interactions) the proton recoils and new particles are also generated; this is "photoproduction"—production of new particles by means of a photon beam. The three spectrometers in End Station A are designed to measure the amount of electrons scattered, the new particles produced, and the recoil of the proton.

Between experimental runs, the usual access to End Station A, unless the walls are open, is through a long corridor opening onto the vast 2,500-square-meter interior space of the building. The most prominent objects in the room are three big concrete boxes, each mounted on one of a concentric set of rails forming nearly a half circle that takes up most of the width of the room. At the center of these rails is the target, a drum-sized metal cylinder mounted on a tall pivot made for a sixteen-inch gun from a World War II battleship. The base of the pivot is in a narrow dark circular pit. The beam from the accelerator comes in along an overhead pipe and meets the target about nine feet above ground. Below that, collars turning on the pivot connect to three bulky, rigid limbs, each one leading its catch of particles from the collisions in the target out to one of the concrete boxes.

The three limbs all look different: each contains magnets that bend and focus the path of electrons and new particles. The amount of resistance to these constraints is the first indication the detectors receive about the momentum and position of the particles at the moment of collision. The two longer limbs are about two meters wide and high; one is about fifty meters long, the other about twenty-five meters. The concrete boxes open to reveal configurations of detectors, which have changed often over the years. Originally, simple counters were arrayed in the boxes. By 1971, these were supplemented by multiwire proportional chambers, so that the data about the scattered and newly produced particles could be further refined. This added information also makes possible better detection and rejection of data from unwanted sources. The multiwire proportional chamber is a new type of spark chamber originally developed at CERN. In the grid used to detect the presence of particles, wires were placed particularly close together, but in a way that avoided the danger of having the whole system trigger itself. Each wire constricted when fired by a passing particle; when that constriction moved (at a predictable rate) to the end of the wire at the edge of the metal frame, the computer was signaled. In this way, a great amount of very precise information was collected.

The longest spectrometer, called the twenty GeV (billion electron volts), is fifty meters long and weighs two thousand tons; the eight GeV is twenty-five meters long and weighs one thousand tons. The third and shortest spectrometer, the 1.6 GeV, weighs 575 tons.[11] The larger spectrometers analyze scattered and newly produced particles; the smallest spectrometer analyzes the recoiled protons. Large electrical cables spew out of the boxes' detectors and are carried alongside the magnets, down the arms of the spectrometers to the control pit. From there, gathered together, they are guided along the floor and up the wall, where they enter the Counting House, End Station A's control room and computer facility.

Each of these three spectrometers is very limited in the characteristics it can identify and the particles it can capture. Particles scatter from the target in all directions; the spectrometers can monitor only those particles whose paths coincide with the three limbs. However, they have two important advantages: first, at 320 pulses per second, an extremely large number of events can be analyzed; second, all three spectrometers can be moved. They can rotate through a total of 165 degrees around the target.

The pervasive grey of the concrete at End Station A (or ESA) and the large, lumbering detector boxes, with their supporting equipment reaching toward the pivot pit, always look to me like great mechanical elephants with their trunks plumbing a watering hole. The room and the spectrometers seem massive, mechanical, and clumsy. It is difficult to remember that this is the fastest detector at SLAC and the site of more than one experiment of major theoretical consequence. The scattering experiments have served to arbitrate between theories on the internal structure of the proton. The design and integration at ESA of the polarized electron beam source called "PEGGY" and a target with polarized protons has enabled experimentalists to design an experiment from which it was concluded that time reversal invariance is maintained—that is, these particle interactions proceed the same whether "time" is going forward or backward.[12]

LASS. Nearby is a building containing another kind of detector, the Large Aperture Solenoid Spectrometer, known by its acronym, LASS. As I described earlier, particle activity in a bubble chamber can be detected in three dimensions because the medium in which the initial collision between accelerated particles and target particles occurs is also the medium of detection. The difficulty with bubble chambers, however, is their slow rate of data collection and the expense of analyzing data from visual records. The hybrid systems minimize these problems, but do not eliminate them. LASS represents another approach.

In an earlier incarnation of this detector, several different kinds of recording devices, including both counters and spark chambers, were aligned behind the target, which is at the core of a very large magnet. As in the bubble chambers, the purpose of the magnet is to alter the path of charged particles and provide a measure of that particle's mass and lifetime. Several characteristics of those fast-moving particles which emerged in a "downstream" direction from the initial collision, continuing the direction of the beam, could be identified. This detector was called the neutral K meson wire chamber facility (the K^0 or the "K-zero"), after the altered particle beam of mesons it used. The group leader was very proud of the degree of resolution achieved in the multiwire proportional chambers. The group leader also noted to me that the construction of these "magnetorestrictive" spark chambers and the Cerenkov counters used with them, which identify particles by their characteristic light emissions, included techniques he had developed as a young physicist.[13]

The K-zero was incorporated into LASS, much as the hydraulic drive system from the eighty-two-inch bubble chamber was incorporated into later bubble chambers built by the same group. The added features were designed to identify the slower-moving particles that had emerged from the initial collision at wider angles than the K-zero could detect. This diverse array of magnets and detecting devices means that a great deal of information can be gathered about each particle path, even though LASS is not a three-dimensional detector. All of the information is recorded by computers, and complex data analysis can be conducted during an experiment. Considerable computing power is required for operating the detector, data collection, and analysis, especially given the speed at which LASS is designed to operate: while the eighty-two-inch chamber recorded 24.3 million events in five-and-a-half years, LASS could record one hundred million events in one year. These massive computing requirements eventually led to the acquisition of a major new computer for SLAC, the IBM 3081.

LASS was designed to investigate interactions analogous to those explored at End Station A. In the ESA scattering experiments, accelerated electrons collided with target protons, generating new particles and causing the electron and proton to recoil. LASS would be able to study the same kind of scattering process, but between different kinds of particles. In particular, LASS would analyze hadron-hadron interactions. Electrons have charge, which means that they belong to a class of particles interacting by means of electromagnetic force. Hadrons are particles that interact by means of the strong nuclear force.

LASS is about twenty meters long, and three to five meters high. Each of the components, aligned in series along the beam path, has a distinctive configuration. The four separate successive coils of the superconducting magnet, three meters in diameter, are smooth, thick, and bright. Sandwiched between these coils are the multiwire proportional chambers; almost all that is visible of them is their long electrical cables. This whole section rests on a system of rails and jacks so that it can be aligned and dismantled easily. All this is straddled by a long-legged rigging, which supports the refrigeration system for cooling the magnets. The rigging looks like a giant daddy longlegs insect embracing a metal and plastic caterpillar.

Next is the conventional magnet. After that the fast-moving particles enter what is essentially the old K-zero detector. Its magnetic coil is wrapped into a configuration that looks like giant lips surrounding the particle path for a meter of its length. Next are the twelve flat, one-meter square wire spark chambers with their mass of electrical cords, hung on bright metal frames. Each frame looks like a vertical loom, and the spark chamber is like a rug with its threads still attached to the frame on all four edges. Seeing a dozen of these frames with chamber and cords attached, arrayed in series, brings to mind a textile mill. The physicists move around and between the frames, adjusting the chambers and tinkering with the cords, like loom operators. The buildings are not heated or well lit, which enhances the image of a nineteenth-century mill. The closely spaced chambers covered with plastic sheeting resemble the bellows of a huge plate camera. At the end of the entire tract is a dark box, five meters long, containing the Cerenkov counters, which extract the final data from the event by analyzing the light emitted by the particles. Hovering over all this, near the ceiling and moving on rails that run along two opposing walls of the building, is a seven-and-a-half-ton-capacity crane. The crane is used to dismantle and rearrange the components of the detector.

LASS took several years to design and construct; the final stages included many setbacks, especially in construction of the magnets. During this time, doubts were raised at the lab and in the larger physics community as a whole about the viability of this detector, and whether the physics it was designed to do would prove worth the resources that had been allocated to it. Some physicists thought that the design was impossibly reliant on engineering precision; others thought that the physics questions being asked were state of the art but that the detector was undistinguished in conception. Doubts like these, I came to realize, beset any project, especially once it is funded. Part of the responsibility of a group leader is to defend the group's project (and budget) against all efforts to diminish it. Years later the technology for on-line data analysis developed for LASS had been incorporated into many other detectors.

Shortly after the LASS magnets began operating, a party was held in its honor. The research group played host in the building housing LASS to all who had contributed, directly or indirectly, to the existence of the detector. The party for LASS included huge kegs of beer; the group was known to have gathered often at a nearby tavern during the long struggles of funding and construction, to discuss physics and their LASS. Each group member could be seen, holding a half-quart paper cup of beer, pointing to parts of LASS and talking animatedly or sitting in the control room bringing up interesting graphics on the video screens about LASS's sensitive self-monitoring system.

SPEAR. North of LASS and beyond the bulwark of End Station A is a facility which almost failed to be built. It is called SPEAR, an acronym for Stanford Positron Electron Asymmetric Rings. The original design called for two pear-shaped rings; the final design has one symmetric ring, but the name remains. SPEAR is one of a growing number of "colliding beam" facilities at high energy physics laboratories, in which two accelerated beams of particles are made to collide with each other, thereby doubling the center-of-mass energy available at the moment of collision. For example, if each beam had an energy of 4.5 GeV, the center-of-mass energy available at the moment of collision would be nine GeV. Collisions between an accelerated beam and a stationary target have a much lower interaction energy, because the recoil of the original particles is so much energy lost. It would take a conventional accelerator capable of accelerating a beam to fifty GeV to produce nine GeV center-of-mass energy. A significant tradeoff, however, is that the accelerated particles actually collide much less often in a colliding beam design than in the conventional configuration, because a beam is less dense than a stationary target.[14]

A special feature of colliding beams is that usually one beam is composed of the antiparticle of the other beam. For example, at SPEAR, one beam is composed of electrons from the linear accelerator. The other beam, also from the accelerator, is composed of antielectrons, called positrons. An antiparticle has most characteristics in common with its particle, but not its electrical charge, which is reversed. A positron is a positively charged electron. Positrons are generated in the accelerator by causing a partially accelerated beam of electrons to strike a tungsten or copper target inserted about one third of the way down the two-mile accelerator.

The resulting gamma rays, striking another target, generate electrons and positrons in pairs. The electrons are pulled aside by means of magnets, leaving a beam of positrons which then are accelerated the remaining length of the accelerator.[15] Colliding beam facilities use particle and antiparticle beams because in particle-antiparticle interactions no matter survives and there is no recoil: both particles are annihilated, with only energy remaining. This energy then re-forms into new particles, some leptons and some hadrons.

The first colliding beam facility was built at the Stanford University High Energy Physics Laboratory (HEPL) in the late 1950s and early 1960s. The leader of the group that designed, constructed, maintains, and does experiments with SPEAR was a member of the Stanford-Princeton team that built the HEPL ring. He joined SLAC in 1963 and submitted his first proposal for a high energy colliding beam facility at SLAC in 1964. That and subsequent proposals were rejected until 1970, when a relatively modest version was accepted; it still seemed to many a wildly implausible concept. Actually, SPEAR was not funded through the usual channels. The Atomic Energy Commission, the federal agency that then funded particle physics research in the United States, merely allowed SLAC's director to bypass his own Program Advisory Committee and shift five million dollars from other projects to finance the construction of SPEAR.

The man leading the SPEAR group had done important work in the 1950s on electron-positron interactions, establishing that the current analysis of the electromagnetic force was correct at extremely short distances (10^{-13} centimeters). He had decided he wanted to study hadrons, which interact by means of the strong nuclear force:

> It seemed to me that the electron-positron system, which allowed one to produce these particles in a particularly simple initial state, was the right way to do it . . . That was the beginning of the long struggle to obtain funding for the device.[16]

In spite of the precariousness of funding worldwide for high energy physics since the early 1970s and the difficulty of funding SPEAR, a report by this group circulated in 1972 at SLAC projected the following timetable:

> In 1972—conceptual design and physics studies; 1973—begin detailed engineering studies and writing of proposals; 1974—submit official proposal to the AEC, asking authorization for construction to begin in FY 1976. With such a timetable, a PEP accelerator [a much expanded colliding beam facility on the model of SPEAR] would be operating about 1980.[17]

With small perturbations, this timetable's predictions were fulfilled. It is typical for a lab's new projects to enter the design stage immediately after other projects have been funded, so that there is a continuous cycle of ideas seeking funding. Of course, many projects never make it beyond the design and proposal stage.

The main approach to SPEAR is by a narrow bridge over the ring, which is about seventy-five meters across. To a visitor driving over the makeshift bridge into the circular paved parking area occupying most of the middle of the ring, SPEAR looks like a series of concrete boxes closely arranged in a circle. At the edge of the parking area is a corrugated metal building; scattered around the paved lot are some metal sheds for supplies and a trailer used for office space. Interrupting the ring at opposite sides are two more corrugated metal buildings. Walking inside one of these two buildings, one is confronted by a deep, paved rectangular pit surrounded by a narrow walkway with a metal railing. Sunk halfway into the pit is a 150-ton octagonal object standing on edge, fifteen feet thick and fifteen feet high, with people crawling around it and inside it. This is one of the SPEAR detectors, called MARK I. It looks like a huge mechanical model of an eye, with a hole for the lens and eight concentric rings of alternating grey metal and black plastic for the iris, all outlined by black metal. Masses of little wires link the rings. Radiating from the outer ring are about twenty-five light grey metal rods, alternating with pairs of protruding tubes. Thick wires extend out of two sockets sunk into each of those tubes and converge into fat electrical cables. The rods and the tubes are framed by eight grey metal bars. Each bar is echoed by two behind it. These bars look like retracted eyelids, and the cables drape over them like tangled eyelashes.

What we are seeing is a series of concentric wire chambers, sheaves of tubular scintillation counters wrapped in a magnet, all surrounded by more counters. The hole in the center is for the flattened ten-by-two-inch pipe, which carries the two beams of electrons and positrons traveling in opposite directions around the ring. When the beams are at the appropriate energy and density, they are deflected toward each other while they are passing through the section of beam pipe that is surrounded for fifteen feet by this detector. Paths of the particles created from the energy of the electron-positron annihilation are recorded as they scatter through the detector.

Retracing our steps to the parking lot, we enter the metal building housing the Sigma 5 computer that runs SPEAR and MARK and collects and analyzes data from the experiments. Watching the computer graphic display on CRT screens, one can again imagine that one is actually seeing the tracks of particles created from the annihilation. It is a rather tame experience, compared to standing on top of a shaking bubble chamber, but in fact the spectacle is less predictable: within three years after SPEAR and MARK I were completed, two entirely new particles and several resonances were discovered.[18] The weekend that the first new particle manifested itself, a lot of champagne was drunk in the control room at SPEAR, and many people crowded in to see the data reconstructed on the CRT. This experimental result confounded some existing theories and led to a Nobel Prize for the group's leader in 1976.[19] The prize was shared with another physicist at Brookhaven National Laboratory (BNL), near New York City on Long Island, who produced similar results, using a different type of detector, almost simultaneously. The working styles of the two groups were considered by physicists to be diametrically opposed: the SPEAR group did "horseback" physics—aggressive and daring, with charismatic leadership; the BNL group was said to be "finely tuned," meticulous and cautious—and its leadership was known as fiercely authoritarian.

The different names that the two separate groups gave to the same particle reveal their claims of ownership. The SPEAR group at SLAC called the particle *psi* because the highly prized computer graphics of the data produced by their detector generated a visual pattern that strongly resembled the shape of the Greek letter *psi*. Calling the particle *psi* called attention to their distinctive data collection process. The Brookhaven group named the particle J because that capital letter in the Roman alphabet strongly resembles the Chinese character of the group leader's name. In one case, the name evokes nature's presumed signature as revealed by a specific detector; in the other, the name evokes the signature of a human being. The perpetual repetition of these signatures in the data continually reiterates both groups' claims to its ownership.

Gossip was intense: exactly how could the two groups have made the same discovery at the same time? Attention focused on "leaks"; some were sure that rivals of either the SLAC or BNL leader had told former students or research associates at the other lab where to look for the important new data. Many were startled that groups with such different styles should have produced the same results independently. The scientists' disbelief contrasts sharply with the response of the lay person, for whom recurring instances of simultaneous discovery seem to endorse the truth of scientific results.

There was surprise, too, at the awarding of a prize for such recent work; some physicists at the lab said that, since funding in high energy physics had declined drastically worldwide, the Nobel committee had been urged to make an award in particle physics soon, in the hopes that funding would be stimulated. In both these cases physicists responded with a mass of hypotheses to what struck them as anomalous information: "simultaneous" discovery by two groups having disparate styles of physics, and early award of a Nobel prize.

The physics community also responded with a mass of hypotheses to the data generated by the experiments at SPEAR and BNL. There were so many interpretations that *Physical Review Letters* declined to publish them for a period. "Phys. Rev. Letters," as it is called, publishes brief articles on important issues rapidly, on the presumption that more expanded work will be published later in *Physical Review*.[20] A satire on the proliferation of theories to account for the new particle was submitted to *Physical Review Letters* by Marty Einhorn and Chris Quigg of Fermilab, who named their own theory "Pandemonium."[21]

The KEK detector. Soon after my arrival at KEK in summer 1976, a Japanese physicist whom I had met at SLAC offered to show me the accelerator and the research areas. The accelerator is only a short walk from the main laboratory building. The physicist showing me the accelerator explained the beam characteristics that had already been achieved.[22] I was startled to realize that there were no detectors arranged around the usable beam. It was explained to me that no research group was permitted to use the beam because the accelerator department did not want it used when it was only partially completed. I had not been at any other laboratory in which accelerator physicists had the power to overrule experimentalists, but I had never been at a lab at its inception: I know that accelerator physicists are only in command of a laboratory until the accelerator is running efficiently; then the experimentalists take over. At KEK, the accelerator physicists had managed to convince the director that theirs was the correct policy. Still, this did not resolve my feeling that here was something deeply strange: I had never seen a laboratory which was not trying to "get physics results" as quickly as possible. I had become attuned to the sense among American particle physicists that time is a precious commodity. I strongly suspected that at SLAC an experimentalist would have argued that he needed to test his detector, and then surreptitiously done physics research. If the physics results were interesting, he would be lauded for his cunning and enterprise.

I asked a Japanese physicist who had worked in the United States for many years and was then visiting various groups in Japan how he would explain this situation. He replied that he was very skeptical about the whole operation of KEK, especially because of the caution about doing physics immediately. I set out to understand why the press of time did not dominate decisions at KEK, to explain why there were no detectors using that beam.

I arranged to meet with a young physicist who I knew had responsibility for an important section of a detector for a powerful research group. I found him in his research area, which was a space about the size of End Station A at SLAC. The room was being used to house a complex array of activities. A trailer was being used for computer equipment. A target stood in one corner. There were storage areas, work areas, and construction areas. Everything looked temporary and makeshift. The young physicist, in jeans, a tank top, and tennis shoes, was in the process of trying to organize and rearrange the area in order to be able to set up the equipment for his detector. He pointed out that the first priority in the laboratory workshop was accelerator development, so it was very difficult for him to get people's work time allotted to his project. He was also having difficulty getting access to the workshop himself. Consequently, he was rearranging the research area so he himself could work there. He was very busy, and eager to get the detector built, but he was working almost alone.

When I spoke to senior physicists, they began to talk of delicate arrangements still under way in the organization of the laboratory.

They also mentioned the severe limitations on the laboratory funding. The absence of detectors seemed to be related not so much to factors within the lab as to KEK's relations to both the government and the rest of the Japanese particle physics community, an issue to which I will return in Chapter 5.

One day a KEK research group leader said he wanted to show me something that was very important for his group. He took me into a room in which a group of young men were arranging lights and cameras in a professional way, preparing to photograph the only other thing in the room: a computer. The physicist explained to me that an international group had recently judged this computer system to be one of the year's one hundred best . . . well, he could not remember exactly best what. Nevertheless, he knew that these one hundred achievements would be exhibited soon at the Chicago Museum of Science and Technology. He searched around and found two eight-and-a-half-by-eleven-inch glossy photographs of the system which he wanted to give me. I asked who designed it. He responded that he and his staff had drawn up the general specifications and then tried to interest a computer company in the project. He said all were reluctant, but Toshiba finally agreed. He smiled, and added that now Toshiba is very glad to have done the project because the publicity will be valuable to them. He expected that Toshiba's award would help him to interest other manufacturing companies to bid on his research equipment contracts in the future.

When I asked about the construction of the KEK bubble chamber, I learned that—as in the case of the computer—the group had made the general design specifications and then looked for manufacturers to produce the components. The body and the supercooling system were built by Nippon Sanso Company, the expansion system by Kayaba Company, the magnet and power supply by Hitachi, the window by Ohara, and the photographic equipment by Canon. Nippon Sanso also took the contract to construct the chamber from these components. This physicist said with a shrug that the process of finding and choosing manufacturers is difficult.

When another physicist at KEK was showing me a special target his group had designed, I asked about the components. He said that some of the parts were from the United States, France, and England. He pointed out that each purchase from a foreign manufacturer must be made through a Japanese representative of that firm, according to government regulations concerning all expenditures at KEK. All funding is "line by line," and closely watched by government bureaucrats; any expenditure over five hundred dollars must be approved by the government. When I commented that this must be very time-consuming, the physicist laughed and said that after a while one learns whom to approach, whom to call about the progress of a request, and so on. With a smile, he added, "It is important, when applying for purchase of duplicate items, not to photocopy old requests, but to have each new request typed nicely." The implication was that the bureaucracy was annoying, but it was possible to learn their habits and humor them.

Yet another physicist told me of an international conference on governmental expenditures for science which he had attended. At this conference, Japan was congratulated for having the most effective science expenditures. He said he thought to himself that the organizers of the conference did not know what they were talking about. In his research group, he estimated that at least 75 percent and probably 90 percent of the budget was paid to private industries for components and fabrication of research equipment for this detector. The government had encouraged the Diet, Japan's parliament, to pass KEK's annual budget, apparently, because of an understanding that a very large proportion of the money would be an indirect means of subsidizing research and development in private industry.

Experimentalists at KEK, conceiving of an experiment, first define their physics question, formulate their hypothesis, and outline how the hypothesis could be tested experimentally. The design of the research equipment is only sketched. Then the engineers from the appropriate industries are invited to join the discussion. When KEK physicists and company engineers have come to an agreement about the general design specifications of the detector, the companies build the detector components and another company assembles the parts at KEK.

This explained another absence I had noticed in 1976 at KEK: the experimentalists did not have the means to make and unmake their own equipment, even if they wanted to. Their own training had not prepared them for dismantling and rearranging detectors, but for overseeing the work of engineers. The funding system in Japan led to large initial outlays of money to construct research equipment, very little subsequent funding for modifying their detectors, and little hope of new funding for decades. This meant that the Japanese physicists were inclined to design all-purpose, durable detectors and to push for the most sophisticated technology available for each component. The lack of workshops and technicians also meant that they wanted reliable detectors. Like a house in Europe, a Japanese detector is expected to outlive its builders.

As I walked around the research areas the few technicians I did see all seemed, compared to those at SLAC, very young—probably under thirty years old. A physicist explained that most of them had been hired directly from technical high schools. When I asked why none had been "raided" from private industry, or from other government laboratories such as the Electrotechnical Laboratory or the Institute for Solid State Physics, the physicist explained that all potential government employees must take a stringent examination, and that this examination is "valid" only for three years; to request another position or transfer, after three years have passed, the applicant must take the examination again. This routine creates an effective barrier against midcareer changes. No such reexaminations are required for academic positions.

Technicians at KEK can be promoted to the position of research associate and even associate professor, but "that gate is very narrow." At KEK in 1976 there was one associate professor (in the accelerator department) who was a former technician, and one research associate (in the bubble chamber group) who was still a technician—he was someone the bubble chamber group had wanted to hire who declined to retake the exam. Hiring him as a research associate circumvented the problem.

Another research group had had the option of hiring a technician or research associate, and chose a research associate because the training of the technician would have required too much of their time. Almost everyone assumes that KEK must hire high school graduates and completely train them to the requirements of research science technology. A leading physicist at KEK asked me many detailed questions about the role and status of technicians at SLAC. He said that in Japan all very good university-trained technicians work in private industry and cannot be lured to KEK because there are no real opportunities for advancement. He believes that KEK will never excel in experimental physics until this situation changes.

I had learned why the KEK experimentalists had not surreptitiously pushed an unfinished but functioning detector in front of a usable beam. They were not only abiding by the policies of the accelerator physicists, they were also dependent upon the companies building their detector, and the companies determined when the equipment was finished and ready to be turned over to the lab. I had also learned why the experimentalists preferred reliable, durable, and technologically sophisticated detectors. The special constraints on their funding, independent of the amount, and the government civil service constraints on their hiring decisions had led them to a certain style of detector design. I was told by a European experimentalist at SLAC that similar constraints exist in some labs there, with similar consequences. Circuitously, their durable precision equipment is consistent with an emphasis on experiments designed to collect precise measurements of phenomena first identified elsewhere, in more innovative and flexible detectors.

There are many ways to design a detector. LASS was designed to provide data for meticulous analysis. It was focused on a specific class of particles, for the purpose of clarifying the fine structure of a sketchily understood part of a widely accepted theory. The detector itself stretched existing technology in order to analyze certain patterns in very large numbers of events. It has strongly influenced subsequent detector design, even though its physics results were not startling. The LASS group's leader said that a detector should be seen not as an end in itself, but as a piece of equipment designed to "resolve a philosophical question." He sees the strength in his group and its detector as the power to correlate data analysis with deep physics questions.

The group at SPEAR, on the other hand, sees itself as having designed its detector architecturally rather than analytically. This group regards its detector and their physics interests as changing rapidly. The detector is designed so that it can be quickly altered: "We're too stupid to build it right from the beginning, but we can build it so that it can be fixed easily." This physicist means that their group was smart enough to build ways of correcting their "error bars" by fixing the machine itself, refining, deconstructing, and reconstructing it as data analysis improved: "If the detector's architecture is good, new parts will fit in." This detector was designed to search for a wide range of unexpected rare particles interspersed in very large numbers of events. Once the researchers found candidate rare events, they needed to be able to eliminate all possibility that the events were not significant, that they were explicable in any alternative terms. The flexible architecture of the detector enabled them to do this, because they could rearrange the detector to eliminate each alternative explanation systematically. This detector too has had much influence on later detectors.

One physicist constrasted the SPEAR approach to that of ESA: "Our detector was built on much less money, and we are better for it: we built with much more thought and ingenuity. Their machine was built in fat times, and you can still see it in their cupboards. If you wanted three of something, the leader said to order a hundred; we will use them eventually." The implication was that the ESA group was just mechanically rebuilding its detector because it had so many spare parts. This third group made no claim to elegant equipment or subtle architecture. It had built its reputation on quickly getting finely structured results to a specific set of questions. The ESA's detector's results are widely regarded as reliable because the machines are considered to be overbuilt, but other experimentalists have not emulated its design in later detectors.

The differences among these detectors serve as a mnemonic device for thinking about the various groups' models for scientific method: how to elicit traces from nature that are both significant and reproducible. Detectors themselves, then, supply a system for classifying modes of discovery. Each is the material embodiment of a research group's version of how to produce and reproduce fine physics, how to gain a place for the group's work in the taxonomy of established knowledge. Each of the groups' strategies for finding traces is in fact a strategy for dealing with noise. The unfinished detector at KEK was designed to minimize noise at the expense of finding new data. LASS is spare and elegant, meant for refining accepted but little-understood knowledge. SPEAR is ingenious architecture, meant for reconstruction and deconstruction. The ESA is fat and overbuilt, meant to be reliable.

In this chapter I have emphasized the differences between the detectors and the styles they represent.[23] It is important to remember that, in spite of these differences, almost all physicists in American labs have worked at several different short-lived detectors in the course of their careers. In fact, the American physicists' careers often recapitulate detector history. At SLAC the eighty-two-inch bubble chamber, ESA, LASS, and SPEAR emerged in that order; many of the physicists at SPEAR originally worked in the ESA group and the leaders of SPEAR and LASS had worked together in the bubble chamber group. The long-term members of the SLAC community see each other daily, attend the same seminars and policy meetings, and compete for the same recruits, accelerator beamtime, and resources to build their next detectors. Their different strategies for discovery have emerged in the context of a shared perception of how to do good physics and how to maintain a strong laboratory. By contrast, most Japanese physicists work at one detector all their careers and have less contact with other experimentalists; they must pass on their very long-lived detectors to new generations who will ask new physics questions. Across the generations they will share an approach to physics and a commitment to a strong research group. As we turn in the next chapter from the detectors to their makers we find that just as there are different strategies for making research equipment, there are also different strategies for building a career in physics.

CHAPTER·3

Pilgrim's Progress: Male Tales Told During a Life in Physics

Like many social groups that do not reproduce themselves biologically, the experimental particle physics community renews itself by training novices. Gradually, the young physicists learn the diverse criteria for a successful career. This transmission of meaning occurs not only in formal education, but also in the daily routines and in "the informal annotations of everyday experience called common sense," including stories told within this almost exclusively male community about important people and events.[1] In this chapter, I tell the tale of a pilgrim's progress in physics. It is by making progress on this journey that the pilgrim becomes a scientist. The journey itself is marked by the telling of moral tales. The five tales I shall recount are about anxiety and time, success and failure, and I shall attempt to explain how these stories could only be men's stories.

There are three stages in the education of American particle physicists: undergraduate training, graduate school, and research associate appointment, which together comprise about fifteen years in a physicist's life.Typically, only the third of these stages is played out within the precincts of a major lab like SLAC or KEK; only at the completion of all three stages does one become a full-fledged member of the particle physics community. Each stage is marked by distinctive intellectual qualities which the novice must display; each stage also cultivates certain emotional states.[2]

Undergraduate physics students, to be successful, must display a high degree of intellectual skill, particularly in analogical thinking.

The students learn from textbooks whose interpretation of physics is not to be challenged; in fact, it is not to be seen as interpretation. They learn to devalue past science because it is thought to provide no significant information about the current canon of physics, but they also learn, from stories in their textbooks, that there is a great gap between the heroes of science and their own limited capacities. It is not until graduate school that avuncular advisors introduce the older students to the particle physics community and cautiously allow them to see themselves as members of it; through stories of success and failure, stories about the work of the generation now in power, they teach the novices a style of "doing" physics and a "nose" for good issues. Graduate students are also learning to be meticulous and very hard working.

Self-assertion and bravado must be added in the third stage, when the "students"—having earned their Ph.D.'s—become research associates. This desired ethos stands in counterpoint to, yet must include, the meticulousness and patience of the earlier phase. The "postdocs" begin to learn and communicate about physics orally, rather than through books and articles. They cultivate competitive and acerbic conversation to display independence and a contempt for mediocrity. They learn stories about those in recent generations who have "made it"; from these stories they realize how important it is to anticipate the future. Only at the completion of these first three stages can a postdoc become a full-fledged member of the particle physics community. Even so, about 75 percent leave the field after this fifteen-year training period.

Each stage has its characteristic anxiety: for the undergraduate it is a fear that one's own capacities are insignificant in comparison with those of "real scientists," or even inadequate to win admission to the community. Graduate students are afraid of using up their predoctoral years working for a team whose experiments may prove unproductive; afraid of losing their chance at success by losing time. Postdocs are looking two or three years ahead, trying to anticipate rewarding questions in physics; they become anxious about the future.

These anxieties do not disappear when the postdoc gains a permanent position; the fears of the full-fledged member of the community combine all those of the novice (fear of the accomplishments of others, fear of losing present time, fear of the future coming too fast), in somewhat revised form. The established physicists are afraid that they will not *continue* making significant contributions, that they and their work will become obsolete. These fourfold anxieties are indeed inwardly experienced, but they have been learned and cultivated in the community. It is, in part, by these anxieties that we can identify these physicists as members of their culture. In the context of a training period, the novices are learning to want passionately to do what they should by associating emotional states with certain activities.

Undergraduate Students: Instruction in the Margins of Physics

Both Japanese and American university physics undergraduates study the same subjects in the same sequence, and all fully trained physicists, regardless of specialization, are, in principle, qualified to teach any undergraduate course in physics. Students are taught to see physics as a ground of fundamental knowledge on which the various subfields have been plotted. It is their task to learn this rich terrain as thoroughly as possible.

Physics is introduced first to the undergraduate in a textbook. The instructors, who are presented as experts on physics (although not specialists in all subfields simultaneously), explain the material in the textbook. Students are given "problem sets" to solve in order to demonstrate their comprehension of the material. In "easy" problems, the students merely "plug" data into the appropriate mathematical formulae. Harder problems require the students either to recognize data in an unfamiliar form and see that it can be analyzed in ways that have already been learned, or to pick out which known formulae will serve to analyze data which are perhaps deceptively familiar. Discussion clarifies how these choices are made correctly, given the students' level of understanding.

Students also learn stereotypic experiments in highly choreographed laboratory courses. They rarely design an experiment themselves; instead they learn the classic steps in executing an experiment properly. They learn how to calculate the predicted range of error in the data, correct for it, and then analyze the data, according to certain conventional models.

There is no debate about alternative interpretations at the same level of analysis. (There are a few exceptions, such as wave versus particle interpretations; even in this case, what students learn is which interpretation is called for in a given context.) Teachers show students how to recognize that a new problem is like this or that familiar problem; in this introduction to the repertoire of soluble problems to be memorized, the student is taught not induction or deduction but analogic thinking.[3]

Quantum Physics, by Eyvind H. Wichmann, is a textbook that many of the postdoctoral researchers interviewed for this study once used as third-year undergraduates. It presents its topics as unfolding logically. The sequence of the presentation, and the use of adverbial phrases ("around the turn of the century," "then," "today," "now," "to date,"), implies that the ideas that constitute the current canon of physics actually emerged chronologically in this "logical" order. The students are being introduced, subtly, to the physics community's image of its history and the place of history in its present. Practicing particle physicists look to the past only for the direct antecedents of the present; they are interested in their immediate predecessors, who can be labeled the heroic discoverers of current, supposedly correct, scientific ideas. Their history of physics is a short hagiography and a list of miracles. It is this history that they teach their students: a set of oral traditions about heroes and antiheroes, detectors, and examples of "good physics judgement." The physicists' knowledge of these legends usually extends back only to their own student days; beyond this, they know only of those predecessors who can be seen as having anticipated current ideas.[4]

In the margins of Wichmann's text, the heroes of particle physics are identified.[5] These scientific giants are shown from the waist up, alone in an office, or in a portrait pose which includes only their heads and shoulders. The rhetoric of these images underlines the outward physical similarity of the men, their apparent social conformity (all but one are wearing jackets and ties), and their freedom from any particular social context (the backgrounds are blank or neutral). In another, more recent, textbook a very different photographic image—that of Richard Feynman grinning, body askew, playing bongo drums—relies heavily upon the more familiar genre for contrast.[6] In the Wichmann text Einstein, in a crumpled sweater, is the only one not wearing a suit.[7] Sommerfeld, in a hat and suit, is the only one shown outdoors (the background is out of focus). The images also reinforce the message that all major scientists are male. Ten are gazing away from the cameras, suggesting internal meditation; five are looking challengingly at the viewer.

A brief caption gives the man's birthdate and place, date of death, universities attended as a student, date of award of the Ph.D., academic positions held, references to early studies if these were not in physics, and the topics of his significant scientific work. Note is also made of any emigration from one country to another.

There is, it seems to me, a cluster of subliminal messages in these picture captions: that science is the product of individual great men; that this product is independent of all social or political contexts; that all knowledge is dependent upon or derivative from physics; that only a very few physicists will be invited into the community of particle physics; and that the boundaries of particle physics are rigidly defined.

Regarding their scientific activity as supranational and supracultural is a way for physicists to isolate their community from conflicts between their countries and maintain the stable communications network necessary for their work. The movement of Germans to Great Britain or Japanese to the United States in certain years suggests that physicists are above politics and nationalism, although political duress can threaten the proper pursuit of physics. Scientists of all nationalities—Swiss, German, Italian, Scottish, English, French, Australian, Polish, American, and Japanese—look as alike as possible in this portrait gallery, signifying that culture is not an issue among them. Listing the names and varied birthplaces again suggests that scientists emerge in all cultures; the implication is that science and an aptitude for it are independent of variables like culture. Most physicists would argue that there are no cultural influences on their activities as scientists.

The universities where the scientists pictured in the textbook have appointments comprise all, and only, the major research centers in particle physics. The schools where these men were educated all provide a good preparation for work in the field, in the judgment of senior scientists. The student is being introduced to the fact that there are at most a dozen major, national and international research laboratories in the world that serve as reference points on the map of particle physics. Physics is an extremely restricted community, and the institution where one practices is the first clue to one's status in the community.

Shifts in fields of study noted in the captions suggest that physics is of more intrinsic interest for great minds than the fields they chose to leave, such as chemistry, engineering, and history. In one book for high school physics students, particle physics is represented as "the spearhead of our penetration into the unknown."[8] (Phallic imagery is found in much of the informal discourse of the male particle physicists. I know of no study of the role of sexual and sexist language in the training of any occupational groups, but such speech is abundantly present in every stage of a scientist's education.)[9] Particle physicists share the assumption that this spearhead has a shaft extending behind it; after chemistry and engineering comes biology, followed perhaps by the social sciences and humanities. They would also agree that the scale of intelligence and of reasoning capacity needed to practice these various specialties corresponds to this sequence, with particle physics the most demanding and humanities the least. Significantly, the fine arts and mathematics are not ranked in this hierarchy of knowledge. Each of them is thought to share with particle physics certain crucial characteristics: art and physics both require creative imagination; exceptional rigor in analysis is needed in both physics and mathematics. Thus, particle physics is presumed to include what is best about art and mathematics, while excluding the rest. The boundaries are finely drawn. Theoretical physicists may be chastised by their peers for being "too mathematical," or, alternatively, for lacking "physical intuition," or "cooking up schemes out of air." Experimentalists must guard against being seen as routinized "engineers" or, on the other hand, as preoccupied with bravura innovations in machine-craft.

Students also learn in Wichmann's margins what is to be excluded from their field of scientific study. One figure in the textbook shows a simple electric motor. The caption asserts that although its operation is to be explained in terms of quantum mechanics by a particle physicist, its design and construction should be in terms of classical electromagnetism and classical mechanics—the domain of engineers. The caption of another figure implies that the commercialized, popular understanding of science is merely fanciful and bears no relation to the proper study of particle physics. Women are caricatured physically and intellectually. A figure entitled "The Linear Scale of Things" graphically ranks the vast domain of physical things dependent upon elementary particles and their interactions. There is an inverse relation in the things represented between size and seriousness; the upper end of the scale is exemplified by "cute" and/or irritating creatures, from nude females to fleas. Undergraduate physics students are being asked to shift their attention from what is visible and emotionally engaging to the lower end of the scale, which the text proposes is fundamental, where nature is no longer accessible to the naked or optically aided eye.

Beyond these messages in the margin, there are instructions for the students in the body of the text, about their own status as novice physicists. They learn that information taught at each stage is often distorted or partial, a very rough approximation of the truth, which is to be disclosed at later stages. Novices are thought to be unsuited to a full disclosure of truth in these first years.[10] Stephen Brush has used textbooks from the Berkeley Physics Course Series to show that the students are urged to assume that they are not going to be an Einstein or Dirac, but merely soldiers in the ranks who must learn the established rules for puzzle solving within the existing theories.[11] Recall that Francis Bacon in the *New Organon* advocated his method of investigation precisely because it enabled even the most conventional minds to make solid contributions.[12] According to the Wichmann text, the gap between the students and the heroes identified in the margins is quite large.

These pictures and captions are consistent with what Brush describes as the "public image of scientists as rational, open minded investigators, proceeding methodically, grounded incontrovertibly in the outcome of controlled experiments, and seeking objectively for the truth."[13] According to this idealized image of science and the scientist, rewards in the form of prizes, appointments, and publication are bestowed on those who conform to an ethos of science which values, in Robert Merton's terms, "humility, universalism, organized skepticism, disinterestedness, communism of intellectual property, originality, rationality, and individualism"—incidentally implying that these values are not incompatible.[14] A senior physicist at SLAC told me he thinks Merton's description corresponds to "an adolescent fantasy which keeps the students working" through their graduate school years and into the postdoctoral period, when they begin to get some rewards. A sociologist has reported that one of his respondents (a scientist) "indicated that the only people who took the idea of the purely objective scientist literally and seriously were the general public or the beginning science student."[15] Brush also finds that this rhetoric of disinterestedness is used explicitly to inspire young students to pursue a career in science.[16]

These textbook images of the scientist-hero, his personality and his exploits, have much in common with the fictional mode of romance. The narrative mode will change as the neophyte advances in training to become a physicist; but the image of the scientist as romantic hero has a significant place in shaping the first phase of initiation. Northrop Frye characterizes the romantic hero as a most exceptional person whose actions are marvelous; who, with great tenacity and perception, reorders our understanding of the laws of nature.[17] Japanese physicists urged me to remember the scientists in science fiction films both Eastern and Western: against great pressure to conform to the prevailing view, a young man holds to his belief in science with strong conviction. When humanity suddenly faces great danger, it is the scientist who alters the people's perception of the event, provides a solution, and thereby enables the threat to be controlled.

Graduate Students

During graduate school, the physics students are separated into subfields (solid-state physics, particle physics, plasma physics, astrophysics, and so on). Among the particle physicists, the potential theorists and experimentalists are in the same classes and mingle easily with one another. In the classroom, information is presented both through textbooks, which are essentially didactic, and through reprints of articles (papers or letters from journals). The teacher is a specialist in the subfield; the purpose of discussion continues to be clarification of the tasks to be mastered, rather than interpretation. In one graduate-level course in physics, the faculty member asked the students to present papers on special topics in the field. All the students were startled; they had never actively contributed to a course. Some students complained to the department chairperson; they believed that the teacher was not fulfilling his responsibility to teach, and they thought they would not learn much of value from the presentations made by their peers. Students seem to have acquired a model of knowledge as a commodity: the aim is to accumulate as many facts as possible.

In the laboratory graduate students are usually given routine tasks such as dismantling, repairing, and rebuilding a piece of malfunctioning equipment. In effect, the novices are to validate their understanding of the original by making a model of it. In Balinese painting, it is a test of competence for the artist to produce a uniform background of fine detail; it is only in this context that a dynamic foreground can be seen as significant information, as deliberate and achieved rather than as error: "the function and necessity of the first level control is precisely to make the second level possible."[18] The experimentalist calls this first level of control in physics a "coherent ground state"; students must show that they can comprehend and reproduce this degree of order.

In generating a coherent ground state, the students also learn how to differentiate between errors and significant deviations in their data, and come to understand the difference between mediocre and good experimental work. They are learning to become meticulous, patient, and persistent, and that these emotional qualities are crucial for doing good physics. They also are beginning to learn what is meant by "good taste," "good judgment," and "creative work" in physics.[19] They are receiving training in aesthetic judgments as well as in the emotional responses appropriate to those judgments (catharsis, pride, satisfaction, pleasure).[20] They are learning to live and feel physics. Most physicists said that it was in graduate school that they got their first "real feeling for physics."

During graduate school, the student either chooses or is assigned to an advisor. Typically, this person assumes an avuncular role in the novice's life.[21] No matter how outrageous or dangerous the student's errors in the laboratory, the advisor is obliged to remain tolerant. Advisors occasionally take their students "for a few beers"; the advisor tells stories and the novices joke with him.[22] During laboratory meetings, the students may briefly, teasingly "heckle" the advisor. When the student finishes graduate school, the advisor is expected to make use of his network to find the student a postdoc position. Senior experimentalists often comment on the pervasive impact their advisors had on both their personal and professional lives. Most say that they got their "sense of how to do good physics" from their advisor. In the words of Roland Barthes, "the origin of work is not in the first influence, it is in the first posture: one copies a role, then by metonomy, an art; I begin by reproducing the person I want to be."[23]

The laudatory phrases I found most frequently in a group of letters of recommendation for young physicists about to finish graduate school suggest that the ideal student is a "hard and willing worker," "careful, meticulous, and thorough," a "good colleague" who always "delivered" and had a good sense of what was possible.[24] A postdoctoral researcher at SLAC (at the large research laboratories graduate students usually report to postdocs) said that graduate students "should do what they are told, work hard, and not ask questions." Older physicists complain about the liability of having graduate students sojourn in their groups: "They are too much trouble; you must always watch them." "You must check everything that they do, expect mistakes, and just hope that they don't kill themselves [around the sensitive and dangerous equipment]." The novice is learning by a process of trial, error, and comparison.[25]

Two of the letters of recommendation mentioned the wife of the candidate, suggesting that she "understood" that it was necessary for an experimentalist to spend many nights with the detector. Anecdotal evidence suggests that high energy physicists typically marry as graduate students, and rarely divorce. In my discussions with about fifty wives of high energy physicists almost all were deeply impressed with the value of their husbands' field of work. Several nodded vigorously when one well-educated wife in her late thirties told me she thought it selfish and silly for a high energy physicist's wife to pursue her own career. She thought that one could best contribute to society and civilization by providing as much support as possible for the work of people like her husband; very little else could be as important. The older wife of a laboratory director interjected that she thought the way to avoid distracting the husband from his important work was to pursue one's own interests seriously. By their glances at each other I gathered that the younger women disagreed.

In the physics community as a whole, the proportion of wives who strongly oppose making a serious commitment to challenging work of their own is very large. In contrast, the wives of almost all of the most successful senior physicists have developed and maintained strong careers in addition to raising children (I saw men engage in parenting very rarely). This difference may be due to many factors. I suggest three possibilities: that those physicists destined to become distinguished are more likely to marry women who will sustain active careers; that the career interests of wives emerge if their husbands become especially successful, perhaps even as an emblem of that success; or that historical circumstances affect marriage values: those who married during World War II, for example, may have regarded a woman's having strong interests of her own as more acceptable than those who married in the 1950s, 1960s, and even the 1970s. Even among this group, however, the husband's career in physics had always taken precedence in decisions about where the family should live. I found only three dual-career couples (in which both careers were apparently given equal weight) in my entire research on this community; in two of those three cases both husband and wife were high energy physicists.

In general, the massive level of support expected by those engaged in scientific research extends not only to the laboratory staff but also to their families.[26] The families are part of the culture of high energy physics and the stages of a career in physics mold the family, much as one might expect in other vocations such as the military, religious groups, and the arts. Those who are not married by the end of graduate school or those who are divorced can expect discussion about their personal life to be active and elaborate. I have heard male physicists in the United States, France, Germany, and the Soviet Union, but not Japan, analyze the rumors about the supposed sexual liaisons of unmarried physicists from around the world. Very rarely have I heard male physicists discuss the affairs of their married colleagues. Liaisons are discountenanced as an unworthy distraction of vital energies; from this commentary graduate students learn that a successful physicist is a married physicist. The mother of one physicist said that her son had told her he planned to marry because he did not want to bother with a social life that would "distract from his work."

Graduate students also learn stories about male scientists going to extraordinary lengths to get, record, and save data. One story concerns the bubble chamber at the now-defunct Cambridge Electron Accelerator (CEA). Bubble chambers have very sensitive and very powerful pressurization systems. As the perhaps apocryphal story goes, one night one of the propane tanks exploded, practically blowing the students out of the lab; they could have been killed.

One realized he was going to lose the data for his thesis and ran back in to get it; the second explosion blew him out the door again, data in hand.

These stories follow the literary form not of romance but of high mimesis: the hero is "superior in degree to others, but not to the environment"; he is a powerful leader who is subject to the highly consistent order of nature, not a remaker of it. Whereas the romantic hero is secure in his place in the pantheon, the lesser hero of high mimesis may fall. If the story is a tragedy, his fall does not call heroism or the hero into question: he is then a "strong character in a weak situation."[27] The emotions typically generated by the high mimetic form are pity and fear; the form emerges historically at a time when an "aristocracy is fast losing its effective power, but retaining a good deal of effective prestige."[28] The novice physicists, about to escape the power of their texts and teachers, have already begun to draw their heroes from their own kind.

Postdoctoral Physicists

The postdoctoral period, which can itself last as much as six years, concludes the long apprenticeship. Young physicists will be evaluated during this postdoctoral period to determine who will be invited to join the particle physics core community, who will become a member of one of the peripheral groups, and who will leave the field. The core community is composed of the researchers at the major laboratories and the university physics departments whose faculties include several high energy physicists doing research at the major laboratories. Peripheral to them are the so-called facilities groups at the major labs, which maintain the research equipment, and physics faculties at lesser universities. Those who leave the field usually move into computer science, astrophysics, biophysics, or geophysics.

From the Ph.D. stage on, the successful theorists and experimentalists separate. The theorists join Theory Groups and the experimentalists join Research Groups. The experimentalist postdocs are expected to acquire and exhibit skills in detector design and construction, the design of experiments, and data analysis. Ideally, one should become equally adept at all three; at least one should not reveal a strong distaste for, or incompetence in, any of them.

Versatility is important: it is not good to be identified as either a "desk" physicist or a "floor" physicist.

Postdocs are no longer students but young physicists; as such they must learn how to rely on oral rather than written information. They are expected to begin to move beyond the textbook and literature phase of their training, to scan written material primarily to find out who to talk to. Postdocs also learn about the past, conveyed to them in various oral traditions quite distinct from the history of science they were taught to avoid as undergraduates. They learn that what is current has an expected lifetime of about six months. From tales about earlier vicissitudes of their detector, and the exploits of their group leader in younger days, they learn the lore of local nostalgia.[29] Every troublesome detail is remembered; the detector itself is a mnemonic device. A detector in disuse teaches the fear of the obsolete. Postdocs learn that the living past is shallow; genesis, it is agreed, was about seventy-five years ago; genealogies of their teachers can be traced back around forty years (at least for the most notable scientists). Their broader knowledge of the past extends only to the depth of their own generation. The only ones to bother with the past are either old physicists, who no longer "have any physics left in them," or professional historians of science, who presumably "couldn't make it" as physicists.

Some of the postdocs see oral communications as a subtle tool. A European postdoc at SLAC observed to me that many American postdocs simply don't ask questions—they seem to feel that they might come across as uninformed or even stupid if they did. Another, a non-American and one of the few women in the community—she later left the field—found all this hesitation silly and asked many questions; she says that others privately thanked her for doing so. Yet another postdoc suggested that it was difficult to trust anything one heard because everyone was trying to impress the others by sounding off; he thought it best to learn on one's own. Postdocs are learning to negotiate their relationships with others in their research group, and with the particle physics community as a whole. The sharp distinction between heroes and students is now blurred; the postdocs feel that they have begun to shape their own reputations. Anxiety is no longer focused on their standing with their teachers; they must learn to form their own opinions and acquire new techniques for cultivating the good opinion of their peers. In particular they learn to talk about their own work in ways that will convince others of its significance.

By now the senior physicists expect the research associates to be able to do exacting work carefully, cautiously, and with persistence. Almost all postdocs tell stories of how they produced some unexpectedly essential piece of equipment or software for their research groups at the cost of an enormous amount of tedious labor in the face of serious and very complex—but ultimately uninteresting—obstacles. They usually feel that their product was hastily and greedily incorporated by the group, with little if any acknowledgment or appreciation. They no longer get the recognition they got as graduate students for doing such tasks well. Skill is taken for granted. Stories about this are told with either wry amusement or bitterness. Senior physicists will only say that postdocs must learn to live with such situations.[30]

It is widely understood among the senior group members in American labs that it is not sufficient for a postdoc to do this sort of mundane job well; independent, risky work must be undertaken as well. So far as I know, no one tells this to the postdoc. Some sense it immediately; some discover it when they realize that their hard labor is not getting the attention they want; others never learn.

Independent, risky work can be undertaken only if the postdoc succeeds in gaining sole responsibility for some project, a privilege not readily granted. This opportunity is won by displaying a convincing faith in one's own powers to do a task better than others in the allotted time and within the budget, no matter what the obstacles. This self-assertion and bravado is in part a matter of disdaining the work of others; it is necessary to show that one can and will expose mediocre work, no matter who has done it. In America postdocs learn the appropriate ways to call attention to themselves and away from their peers, and how to call upon senior group members to act as their advocates. A postdoc is expected to display drive, commitment, and charisma.

The desired presentation of self can be characterized as competitive, haughty, and superficially nonconformist. Two prominent senior physicists joked that one should "behave British and think Yiddish, but not the other way around." One group leader said that to convince others of the validity of one's work one had to have great confidence and be very "aggressive"; he added that one needed a certain "son-of-a-bitchness." An employee at SLAC who finds the habits of the physicists very interesting said that in his twenty years of watching them he had concluded that only the "blunt, bright bastards" make it, and pointed out the young physicists at the lab who in his view were clearly "too nice" to become successful. Letters of recommendation for applicants to the postdoc positions often say that even though the candidate is quiet and mild-mannered, s/he does excellent physics; in other words, those are qualities for which a candidate must compensate. Achieving the appropriate ethos can be quite stressful, and this is commented upon by many postdocs. Those unwilling or unable to develop this personal style are not likely to be recognized as serious physicists committed to their work.

In sum, senior physicists in American labs assume that good physics judgment is associated with independence, and they cultivate this trait in their postdocs by erecting barriers to it. The postdocs learn (some gradually, some quickly, some never) that they are actually expected to leap over these obstacles, as they did in producing that essential piece of hardware or software. This situation demands a "careful form of insubordination"[31] on the part of the postdoc: they respect the tacit instructions of their elders by not following their explicit instructions, and this must be done with considerable delicacy.

The postdoc situation is usually described by senior scientists, at any rate in public, rather differently. Paul Berg, in an interview at the time of his 1980 Nobel Prize in Chemistry, commented on how his students and research associates were building their careers by working in his lab:

> My work has been built on the work of many other people, and without the people who worked with me—the post-docs and students—we could never have done what we did.
>
> In fact, many of the useful ideas came from students, while I was away.
>
> When the head of a team like mine gets recognition, everybody in the lab knows that the credit is shared among many. I have not done any research work with my hands for five years. The work may have been initiated or guided by me, but much of the work has been done by students *who are building their careers.*
>
> We work as a unit. Very often we cannot identify the origin of an idea. It will have been adapted and modified and changed so

> often, then it leads to something else, and finally there is a breakthrough.[32]

The discrepancy between the official description of group work as cooperative and the persistent, disguised message that only competition and transgression will prevail sets up a "double bind." Gregory Bateson, the creator of the double bind theory, says of this kind of behavior: "First, . . . severe pain and maladjustment can be induced by putting [one] in the wrong regarding [the] rules for making sense of an important relationship with another . . . And second, . . . if this pathology can be warded off or resisted, the total experience may promote creativity."[33]

Postdocs who discover this double bind too late become angry at what they consider a deception. One group leader implied that those who complain would not have made it anyway when he said that "the unhapppy RAs [postdoctoral research associates] are not the RAs we want."

Two older physicists endorse this ordeal in other terms: "Some do what they say they will do, no matter what; others give you lots of excuses about why it's not possible." "Ordinary ones keep cycling through the same mistakes; good ones become contributors."

There are many stories among postdocs about "making it big," making a major contribution to the group and assuring one's future. As one postdoc says: "There are three kinds of experimentalists: the top five percent, the next fifty percent, and the others. The top ones, no question, they're good. The next fifty; they are very bright, but they need luck and judgement and good opportunities to make it big. And politics. The others don't make it."

A European postdoc said that the Americans are terribly competitive: "They are all trying to make a name for themselves, selling ideas and hiding ideas. They're all looking for their own private quark." An Asian postdoc who had been educated in the United States said:

> People are tense and anxious; it is hard to work with them. It was hard for me at the beginning. People see me as someone from an underdeveloped country who wouldn't know anything about technology. The graduate students wouldn't listen to me. Only after a few months do they begin to realize I know what I'm doing. Everyone has intuitions, hunches about the physics. It comes from experience and knowledge, as yet uncrystallized.

> Once in a situation my boss said to me, "This isn't Eastern mysticism, you know." I don't believe this comes from racism. It comes from the competition. Everyone is looking for some way to make themselves look better, to make the other guy look worse.

Another postdoc knows that he has not "made it," and he is angry. Reexamining his postdoctorate, he believes that he now understands why he failed. In the first year, he thinks, the senior experimentalists are scanning the postdocs to see who is "charismatic." They watch how the postdocs handle conversations; the preferred style is confident, aggressive, and even abrasive if one suspects that another's ideas are wrong; he feels he could have adopted this style if he had understood its importance. The next step would have been to seek out "an action sector"—an exciting, volatile, and fashionable area. At that point in his own career, "charm" became fashionable. He feels he ought to have taught himself and then taught others about "charm." He should have anticipated where the problems lay and cultivated connections accordingly. The next step would have been to gain responsibility for some large, important project—so important that others would have sought him out to talk. Ideally, he should have proposed an experiment of his own. Instead, out of loyalty and commitment, he stuck with his assigned task. Now he feels this cost him his career in high energy physics. He believes that he would have had difficulty being granted these responsibilities, however; he sees himself as being outside the "old boys' club," because in the labs where he did graduate work his undergraduate school was not considered to be "on the map." This postdoc has come to these conclusions retroactively; he is comparing his career over the past several years with those of his peers who have made it. He has since left high energy physics.

The graduate student's fear settles into the misery of the postdoc double bind; that misery pushes a few past their fear into a bravura performance and success. A senior experimentalist told me that someone with experience can judge new postdocs within a few days of their arrival at the laboratory. He noted that it is important to see how the postdoc gets along with others, and that takes more time—"but if he is good, that is self-validating." He did add that sometimes this is not a question of talent, but environment. "It's like bringing wildflowers from Alaska to California. They just won't take root." If the flower does take root, it may become quite valuable to the leader in a few years. Senior physicists believe that it is not necessary to scan the hinterlands for initiates; they are sure that the exceptional candidates will eventually be brought to their attention, whether they come from Caltech and Tokyo University or from the universities of Alaska and Hokkaido. As the Americans say, "If they are good, they will get here."

The physicists see themselves as an elite whose membership is determined solely by scientific merit.[34] The assumption is that everyone has a fair start. This is underscored by the rigorously informal dress code, the similarity of their offices, and the "first naming" practiced in the community. Competitive individualism is considered both just and effective:[35] the hierarchy is seen as a meritocracy which produces fine physics. American physicists, however, emphasize that science is not democratic: decisions about scientific purposes should not be made by majority rule within the community, nor should there be equal access to a lab's resources. On both these issues, most Japanese physicists assume the opposite.

This composite account of the postdoc's fortunes takes the literary form of low mimesis: the suffering hero strives for advancement, which is blocked by seemingly intractable obstacles. The obstacles are finally overcome by an opportunistic and vigorous response to unforeseen circumstances. While such "heroes" can inspire sympathy or scorn, we are not inclined to see them as superior beings, much less as superhuman,[36] a key difference between hero-scientists of the past and of the present.

Paradoxically, to be fully conscious of the social and psychological forces at work in this postdoctoral phase would be debilitating for the candidate, according to this community's values. "Unconscious" in this community means arbitrary and unknowable, and hence uninteresting. Concern with these and related matters, such as how to get along with other people, is considered somewhat unscientist-like. Social eccentricity and childlike egoism are cultivated displays of commitment to rationality, objectivity, and science. Young scientists often assert their ignorance of human motives, of everything "subjective," as if that confirms their vocation.[37] Development of insight into one's own motives and actions is thought to be a diversion of time and attention better spent on science. One experimentalist has told me that he believed a successful postdoc had to be rather immature: a mature person would have too much difficulty accepting the training without question and limiting doubts to a prescribed sphere. He felt that this precondition kept most women and minorities from doing well: their social experience had taught them to doubt authority only too thoroughly.

At the end of their fifteen-year training period young particle physicists hope to become full members of the research community. However, only about a fourth of the American students will be sponsored for positions at major universities or laboratories where they will be able to continue research. Those remaining have three options, which like the less successful Ph.D.'s they take in approximately equal numbers: they can leave the field, or work at peripheral schools and cease research, or take a staff position at a major lab in which they will manage the production and maintenance of detectors (what one group leader calls "engineering management"). During their postdoctoral assignments the young physicists have learned the meaning of these four kinds of careers, as seen by the group leaders, and most have come to believe that this ranking correlates with personal ability. They know that their group leaders' evaluation of them is critical for their futures; they usually believe in its justice.

There are a very limited number of laboratories and universities "on the map" in physics. The novice physicist can learn the names of these "holy places" in physics simply by noting the places mentioned in the biographies of the great physicists located in the margins of the undergraduate textbook. As in the rest of academia and other preindustrial institutions, such as the Catholic church and the military, they are trained at the "core" and then move to the "periphery." While it is possible to move from a peripheral undergraduate or even graduate school to a postdoctoral appointment at a core lab or university, I know of no cases in which a particle physicist moved from a postdoc at a peripheral place to an established position at a core institution. To gain those positions, one needs the support of a powerful group leader with a strong network.

By learning to read the margins the novices learn to become competent practitioners of the culture of the particle physics community. Graduate students learn the boundaries of their community and, as they say, the "fine structure" of the differences within it; later, as postdocs they learn how to comport themselves in situations which reflect these differences. At the major labs, they learn that outsiders are devalued and exactly how this is done and what justifications are given. Work unlike that done at the major centers of research is easily labeled peripheral; difference within the community becomes redefined as eccentricity or deviance. That is, they are learning the role of outsiders from insiders. They learn in detail the different ways of being an outsider available to them as career choices. In short, they learn the fine structure of the hierarchy in the particle physics community before they learn their final place in it.

Clear boundary markers define the community by defining specifically *both* what the group wishes to include *and* what it excludes. In the analytic language developed by Gregory Bateson, the matter in the textbook margins serves as "context markers"—signs that operate at a "higher logical type" than the (social) text—indicating to those who know the cultural clues in the context markers how to read the messages in the (social) text.[38] In other words, the immortal heroes of science in the margins of the undergraduate physics textbook define the posture one must display and the genealogy one must acquire.

Having gained a position at a major institution with the help of their former group leader, young physicists take up their responsibilities in a network and begin to establish their own reputations and build personal biographies to match those in the margins of their undergraduate textbook.

To become a designer of machines one must survive all these exclusionary cuts, show oneself to be "part of the signal, not part of the background noise." In their struggle for survival novices learn about time. During the undergraduate years, students discover the insignificance of the past in high energy physics while simultaneously learning to see the heroes of the past as inaccessibly great. Interest in the formal history of the field, full as it is of outmoded or erroneous ideas, is considered debilitating. Study of the history of physics is condensed into the generic celebration of timeless genius and "reproducing" the successful experiments under descriptions that assimilate them to current knowledge. Graduate students learn to fear losing data by accidental erasure of computer records or failure of detectors during an experiment, which results in a loss of beamtime and hence a lower production of data. Less data means a lower-quality thesis, which means less opportunity for a good postdoc position. Postdoctoral research associates, from stories about those who have made it, realize how important it is to anticipate future new directions in theory and new solutions to the design problems of detectors. In order to be successful, they must privately figure out the future of physics. When the novices become members of an experimental research group, they begin to identify their own careers with that of a detector. After their ten years of training, they now have about ten more years to make their reputations in the field. It is as new group members that the young physicists learn the significance of the lifetimes of detectors, research groups, laboratories, careers, and ideas. Fear of obsolescence in these five areas leads to a recognition that uptime, downtime, and beamtime are scarce commodities to be acquired and used in the contest for power.

Group Leader

During his career, a group leader accumulates considerable wealth in the form of detectors, targets, and computer software, as well as his less tangible—but perhaps even more significant—reputation in the community. (In Japan this wealth is controlled by the entire *koza*: the full professor, the associate professors, and the assistant professors.) That reputation is the power that the leader wields in the community as a whole; it is symbolized by his membership on laboratory program advisory committees, which determine which experiments will be accepted, and by his control of a network cutting across laboratories and physics departments around the world. This wealth must be maintained assiduously; it is not clear that it can be inherited. Every powerful senior physicist can invoke a lineage of which he is a part, naming his teacher and his teacher's teacher. It is direct descent from one leader to another in this lineage and the attendant privileges, rights, and duties that a leader will try to bequeath to one of the postdocs who have worked with him. The generations are about fifteen years apart, which means that a group leader would be choosing among approximately twenty-five postdocs for his presumptive heir.[39] But if the group leader began to assert his influence on behalf of a successor, the other group leaders would begin to resist it. It might appear that a group leader could gain concessions and establish his heir, but in trying to do so he would have to spend some of his wealth—thereby weakening his own position, leaving him little to bequeath. Nevertheless, they try. During the old days of funding expansion they could start new groups for their heirs. Now the only group their heirs might lead is their own. Just as this generation's leaders' power reaches its zenith, their capacity for naming the next generation's leaders is lost bcause of declines in funding.

Among the five to ten postdocs the group leader actively supports during their careers at other departments and laboratories, one or two usually have the ambition to achieve the stature of group leader. These postdocs feel that it is better for their careers in the long range to move outside their leader's immediate circle. If he is able to establish himself, usually with the continued implicit support of the group leader, the young physicist will be extending the group leader's reputation, influence, and network, not depleting it. The leader and protégé will help each other in building and maintaining domains in their respective generations. One group leader at SLAC was known to be angry about a contretemps with another leader at SLAC which had resulted in support being withdrawn from one of his experiments. A few years later, when the second group leader appeared to be generating very important data, the first leader is said to have leaked the information to a protégé at another lab, whose group supposedly began to look for the crucial data. When the discovery was announced, the credit had to be shared by the two groups. This quite possibly apocryphal story made sense to its tellers because it included a leader and his loyal protégé assisting each other in their separate domains, each competing with others in his own generation.

To senior American physicists the process of gaining a position in the particle physics community seems largely unchanged since about 1935, with two exceptions. The first problem, as they see it, is grade inflation. They believe that universities have abandoned their responsibility to maintain high standards, so that a "weeding out" now happens at the postdoc level which should have occurred among the undergraduates. As one said, "Any seventeen-year-old who enters Stanford University now can get a Ph.D." He added that when applicants for positions in his group do not meet his standards, he simply hires no one. It is not true, in his opinion, that there are not enough jobs; saying this is just a "crutch" for those who are not good enough, of which "there are far too many." For this reason, he does not find it difficult to tell some postdocs to leave the field. Many other group leaders expressed these sentiments.

The senior physicists feel that they share a strong commitment to physics; they fear that the younger physicists only came into the field for the glamour and excitement. Some are pleased that biology has become fashionable, because the students who follow fashion will not be in physics. Nevertheless, some physicists are very concerned about so many physics students switching to biology.[40] I listened to one professor at a midwestern university explain for nearly an hour to a student that while the financial opportunities were greater in biology, the "real science" always has been in physics.

The second problem, according to the senior physicists, is that funding for particle physics began to diminish around 1970. (Actually only the rate of increase has declined.) They characterize their field until that time as having "grown exponentially."[41] They mean that a sufficiently talented graduate could have expected before 1970 to spend five years in graduate school, five years as a postdoc, five years in a research group, and then to become a group leader. They realize that this is no longer possible. Now new groups are formed only rarely, because of the attrition of funding. This means that talented young physicists, even though ready to be group leaders, must remain in established groups. The senior physicists believe that this situation is hurting particle physics because so few are able to follow a "normal" career path. It appears that group leaderships will become available only when the senior generation (now in their midforties to midfifties) retires. Almost no one will be able to follow in their footsteps for fifteen to twenty years. In the meantime, the senior physicists are unable to place their protégés at the major institutions, stifling the growth of their own influence in their own generation by slowing the growth of their network.

The Americans believe that these two problems—the decline in quality of students and in quantity of funding—are due to forces outside the particle physics community. Nevertheless, they are compelled to cope with the consequences, obliged to spend what they consider an inordinate amount of time attending to the next generation, both in restricting entry into the field and in getting favored candidates established with research groups of their own.

The Japanese also note the same declining rate of growth in their community, but most explain it as a natural consequence of maturity. They point out that during the 1950s and 1960s, high energy physics was a new field in Japan; the practitioners were mostly graduate students and *koshi* (tenured research associates). Some new *koza* (chairs) were established, and the *koza* in low energy physics often filled their graduate-student and *koshi* positions with high energy physicists. During the 1960s and 1970s, these junior staff advanced rapidly to the positions of *jokyoju* (associate professor) and *kyoju* (full professor).

During the 1980s the field has matured, and growth will be more slow. Some departments have expected new *koza* to continue being established at the same rate as in the 1950s and 1960s; they have had a surfeit of particle physicists with Ph.D.'s whom they cannot place in the limited number of *koshi* positions. The result is the emergence of a new position, between the Ph.D. and the tenured *koshi*, called "overdoctors." They are beginning, in other words, to establish a new position somewhat like the American postdoc.[42]

The Japanese chair system of university organization resembles the British, French, and German models.[43] Funding from the government for university education goes not to the institution, but to the *koza*. A *koza* conventionally consists of one professor (*kyoju*), one associate professor (*jokyoju*), one research associate (*koshi*), and two assistants (*joshu*). The professor, associate professor, and research associate are not only all tenured positions, but one is typically automatically promoted to the next higher position when it is vacated. When the *kyoju* retires, the *jokyoju* and *koshi* move up one step, and a new research associate is appointed. This means that in the Japanese physics community, the moment of admission or exclusion is at the end of graduate school, not at the end of the postdoc as it is for Americans. In addition, there is almost no midcareer mobility from one institution to another in the Japanese academic marketplace.[44] The *koza* reinforces the development of lifelong working relationships, in the context of a clearly defined and noncompetitive vertical ranking of the group members.[45]

The responsibilities of a Japanese group leader are quite different from an American. The resources of the *koza* were bequeathed to him; it is his responsibility to maintain those resources, perhaps add to them, and then pass them on to the next leader, the current associate professor. The distribution of the income of the *koza* for research, salaries, visitors, and so on is decided upon by the *koza* leader, in consultation with all the members. The leader also administers the *koza* finances; in America this would be one of the major tasks of a group leader's administrative assistant, a managerial position almost always held by women who are not scientists and who are well versed in institutional regulations and the informal pathways through bureaucratic labyrinths. Since government funding for research in Japan comes directly to the *koza* rather than to laboratories or universities as in the United States, the *koza* leader spends much time in Tokyo maintaining good relations with the bureaucrats of the Ministry of Education, Science, and Culture (Monbusho), who have no training in science and who change positions every few years. This highly prestigious burden of science administration occupies most of the *koza* leader's time, leaving little for research or research planning. Physics leadership for the group often falls to the associate professor and even the research associates, giving them much more influence than their counterparts have in America.

In addition to his external responsibilities the leader also guides the careers of group members. He makes use of his extensive international network to give the younger physicists experience doing collaborative research abroad. He also encourages all the members to develop broad experience in physics research by assigning them different kinds of tasks every few years. He brings midrank members to university, laboratory, ministry, and industry meetings so that they may learn by observation the techniques of negotiation. The leader counsels group members on whatever problems they may be having, including personal ones, and will provide formal introductions to young women of suitable background for the unmarried young physicists. (I have met only one Japanese physicist over thirty-five who is not married.) In short, the leader has a generative, nurturing role; unlike the American group leader he does not evaluate group members who are confronted with a set of provocative double binds, he does not exclude members from the group.

Toward the end of his career the senior physicist may become a laboratory director, although there are a very small number of them. In rare circumstances he may become a dean or university president. Japanese high energy physicists do not have nearly as much power in formulating national science policy as their American counterparts. A Japanese physicist's career is further foreshortened by the fact that at the prestigious public universities retirement age is usually fifty-five. The *koza* may retire into a position at a private university, but he will not command the same resources at all. He does have an option not readily available to his American colleagues: he can become a consultant to the private companies from which he once purchased equipment for his group's experiments.

While the networks among *koza* in Japan are relatively fixed and roles in them are automatically bestowed upon those who hold positions in the various *koza*, the American physicists must cultivate personal networks. Although Americans cannot bestow their networks upon their protégés, they do have the power to establish these young physicists in positions at crucial departments and laboratories in which they can begin to build their own networks. In Japan, there are sharp distinctions between *koza*, with little inter-*koza* mobility; the Americans have only limited intergenerational ties. Competition in Japan is primarily between *koza* networks; in the United States, competition is ultimately confined within generations, between individuals who are struggling to become group leaders.

Senior American physicists see their system as being undermined by limits in funding and lowering student quality, which inhibit the development of new groups and, hence, the emergence of new group leaders. The Japanese system is being challenged by a strong generational rift, which is intensified by the new model of training the next generation offered by KEK.[46] In an effort to keep the American system intact, group leaders attempt to use their power and authority to get their heirs-presumptive established in new groups, but these attempts are thwarted by members of their own generation who are reluctant to see a shift in the existing balance of power. The existing *koza* networks at university departments and institutes are reluctant to see the emergence of new *koza*, which are independent of the traditional network structure.

Full-fledged physicists in America typically tell stories about how good their own work is and how inadequate the work of others in their generation is. For example, one physicist at SLAC said to me that "there is no one 'in-house' at Fermilab who can tie his shoes experimentally." Another physicist said that some well-known experimentalists are consistently wrong, and that their careers are founded only on their personalities. He mentioned a "personal project" of his; over the years he had followed the career of a colleague internationally known as a "brilliant ideas man" and checked out where this man's ideas actually came from. In talking with colleagues across the country, he had yet to find that even one idea had been original. He added that he had found that *one* experimentalist in the field was consistently right: a Nobel Prize winner who "never has an original idea" but who specializes in "shooting down spectacular experiments." He went on to say that this man has organized a group of devoted, willing workers, whom he works very hard, and has built a standard detector very meticulously. If his experiments corroborate the "spectacular experiment," his team's work is regarded as the proof; if they contradict it, then his group gains credit for exposing error. I have been told many stories during my fieldwork by several senior physicists about this man's supposedly egocentric, authoritarian manner and his unimaginative physics.

Physicists also tell stories about the prowess of their own group and use their highly developed rhetorical skills to persuade others to support their work. In ironic reversal of the graduate student's stories about the fear of losing data, the anxieties of the permanent group member are about having data that no one will notice, about everyone paying attention to someone else's data. The end of an active experimental career in physics occurs at about fifty. It is considered inappropriate for someone over fifty to be making discoveries. One physicist asked me, "Sharon, are there any studies done of at what age people make discoveries? My friend and I—this thing that we've come up with [it was a major discovery in the field]—when we were young it would have been the young people making that kind of discovery. Now, why us? If we make this discovery now, young people should have made it four or five years ago." I replied that "Yes, there have been sociological studies of just that subject, and they show that the distribution is over a whole career, that it isn't localized." He responded that the sociologists "Must not have been studying particle physics. Is there any data just on particle physics?" When I said, "Actually, no," he countered, "Well, I'm sure it's true for those other fields, but not ours, and this just shows what I feel, that the young people don't have what it takes these days."

The senior physicists are convinced that the successes of their own generation were based on brash, youthful intelligence, independence, and competition. They wonder if there was something distinctive about their generation, the people who graduated at the end of World War II, but in general they are inclined to believe that the current young generation is not of the same quality as the physicists of the previous fifty or sixty years.

An undergraduate learns to focus on the present, a graduate student discovers that there is not enough time in the present. Postdocs should learn that the future is too short, that they have to anticipate it in order to have enough time. Full-fledged physicists worry about apportioning time between doing physics and going "on the circuit." Senior physicists know that what one needs to be concerned about is obsolescence. Whether for the laboratory, one's detector, one's career, or even one's own ideas, time is running out again, as for the graduate students, but now it is another kind of time. What senior figures need to do before their accomplishments become seen as the last generation's, rather than the last year's, is to make a transition into being one of the statesmen of the field.

Statesmen in Physics

The first stage in becoming a science statesman is administering a laboratory. Each laboratory is thought by physicists to reflect the necessarily powerful personality of its director. Accelerators come into being because of the vision, creativity, and tenacity of an individual who can gather about him a team of gifted people whose work he directs and coordinates by means of his example, will, and—some would say—whim. In the United States, rule through a formal organization structure is considered bad for physics.

When I ask these statesmen if they still think about experiments, they say, "Well, I'm really too busy for that sort of thing, I don't have time. Of course, I'm interested, I keep up with things, go to conferences." Often at the very top conferences a lot of senior people sit around in their shirt sleeves, clearly enjoying the debate, but they are not the speakers, usually. That would be inappropriate.

Statesmen no longer actually do physics; they recruit students. They get money for the lab; they get money for science, they attend to the public understanding of science. It is utterly inappropriate for junior persons to be doing any of these things. Only a senior person can do this. Furthermore, only a senior person who has made a significant discovery can do it, one who has manifestly been able to avoid for an entire career the corrupting enticements of extrascientific power. On the other hand, interacting with people outside is itself a kind of corruption.

Teaching, administration, and consulting for the government are potentially contaminating because they require the cultivation of skills not thought to be based on reason—in particular, the power of persuasion. The scientists themselves usually claim that they no longer do research because they have no time; other scientists believe that these science-statesmen chose their new role because they no longer had "any science left in them." These statesmen regularly transgress the boundary between the domain of rational laws of science and the arbitrary laws of humanity, but not with impunity. As emissaries to the world of the merely human, they are disbarred from practicing science. In a final twist of irony, they take their place in the margins of the textbooks of undergraduate physics students, heroically guarding the boundaries of physics.

Geniuses

The textbook I reviewed at the beginning of this chapter emphasized the gap between the reading student and the very distant geniuses of science. The second story, about the exploding laboratory, suddenly situated the graduate student novice in physics as hero, gallantly rescuing data for science. The postdoc's story was about commitment, about tenacity in the face of nearly insurmountable obstacles, and about the courage to gamble, and the story of the group leader says that by the end of an exemplary career a good scientist is a negotiator, a talent broker, and a fundraiser. Finally, the scientist may become the subject of textbook stories, a genius of science held in awe by those reading their first stories in science. Those stories of timeless genius regenerate the romance of science. Eminent scientists also generate their own stories about their discoveries. I quote from one of those autobiographical accounts in a Nobel lecture:

> That was the beginning, and the idea seemed so obvious to me and so elegant that I fell deeply in love with it. And, like falling in love with a woman, it is only possible if you do not know much

> about her, so you cannot see her faults. The faults will become apparent later, but after the love is strong enough to hold you to her. So, I was held to this theory, in spite of all difficulties, by my youthful enthusiasm . . . So what happened to the old theory that I fell in love with as a youth? Well, I would say it's become an old lady, who has very little that's attractive left in her, and the young today will not have their hearts pound when they look at her anymore. But, we can say the best we can for any old woman, that she has been a very good mother and has given birth to some very good children. And I thank the Swedish Academy of Sciences for complimenting one of them. Thank you.[47]

Another Nobel laureate concluded a brief scientific autobiography by explaining his feelings for the object of his prize-winning studies: "Writing this brief biography has made me realize what a long love affair I have had with the electron. Like most love affairs, it has had its ups and downs, but for me the joys have far outweighed the frustrations."[48] Such stories express the scientists' deep desire for knowing about nature and their deep desire for acquiring data. At the same time, they also point to what these physicists feel about knowing and think about loving.[49]

In these autobiographical statements nature and ideas about nature are coalesced, anthropomorphized into a singular female love object. The image of real female human beings held by almost all these male scientists is that women are more passive, less aggressive than men. This socially constructed gender difference is used by many scientists to define the relation between themselves and their love object. The scientist is persistent, dominant, and aggressive, ultimately penetrating the corpus of secrets mysteriously concealed by a passive, albeit elusive nature. The female exists in these stories only as an object for a man to love, unveil, and know.

In their careers, physicists journey from romantic readings of others' lives, through handing on mimetic tales of heroic action and quests for survival, to becoming skilled practitioners of gossip and rhetoric. They complete the circle by telling erotic tales about physics, tales transformed into romance for the next generation of neophytes. These stories reflect how physicists come to care passionately about who they are and what they are doing. Together they form a picaresque cycle, which chronicles a journey that begins necessarily with innocence and reports its loss; it depicts the growth of strength and the pain of betrayal, hails the achievement of success, recognizes the signs of grace in eminent discoveries, looks back in erotic nostalgia, and lastly eulogizes the heroic dead.[50] In Western culture the picaresque genre is usually reserved for stories about men, not women: women are not seen as gaining strength and wisdom through the rambunctious loss of their "innocence." The very form of these exemplary tales excludes women as their proper subject.

I am not suggesting that only biological males can participate in the cycle. I am claiming that in this cycle a certain cluster of characteristics is associated with success, a cluster that is part of our culture's social construction of male gender.[51] These stories about a life in physics define virtue as independence in defining goals, deliberate and shrewd cultivation of varied experience, and fierce competition with peers in the race for discoveries. Independence, experience, competition, and individual victories are strongly associated with male socialization in our culture. By contrast, recent studies in Japan suggest that these are the qualities associated with professionally active women, not men. Women are seen as not sufficiently schooled in the masculine virtues of interdependence, in the effective organization of teamwork and camaraderie, commitment to working in one team in order to complete a complex task successfully and consulting with group members in decision making, and the capacity to nurture the newer group members in developing these skills. It would appear that there is nothing consistent cross-culturally in the content of the virtues associated with success. We do see that the virtues of success, whatever their content, are associated with men.

In this chapter I have looked at emotional expression as socially produced, socially constructed, not as private, individual, idiosyncratic interior states alone. There may be some pan-human emotional response to certain situations, but I would say that in any particular culture emotional responses in general are shaped and voiced by that culture in that culture's language. This study explores those parts of emotions which are culturally constructed experiences, culturally named and defined. I examine what are considered acceptable ways of expressing sensations and feelings, and what are considered unacceptable ways. I look at how those states are cultivated, or underscored, or suppressed and what kinds of emotions are encouraged and discouraged.

Similarly, I regard gender as socially constructed, not as a biological fact. That is, I study what this group considers masculine and feminine, male and female, and their normal (and abnormal) relationship of men and women. By examining affect and gender in the stories physicists tell I am exploring part of a larger tale. The informal stories people tell in the laboratory can give us a special perspective on the dominant models of success and failure in a community; those models are not gender-free in form or content. Affect and gender are significant components in the division of labor in laboratory research, as well as in decision making, dispute making, and leadership styles that are part of the whole realm of power and tradition in scientific research. My purpose is not to make a psychological exposé, but to clarify the patterns in everyday laboratory practices.

CHAPTER · 4

Ground States: Distinctions and the Ties That Bind

In addition to the powerful constraints shaping a career in high energy physics—the "rules" that govern the passage from one stage of a career to the next—there are several stabilizing characteristics of this community that affect all members at all stages. These are the beliefs and actions, practiced daily, that physicists regard as sensible, obvious, and true. This shared understanding of their world includes knowing how to make an important set of distinctions in the community and how to participate in networks cross-cutting the distinctions. It also includes a way of thinking which for the members of this community is the only way to think realistically and rationally.

Networks

Networks are the set of relationships that bind the particle physics community. Through those relationships, graduate students are placed, physics experiments are evaluated, and long-range goals are debated and determined. These networks intersect with the formal organization of laboratories and national physics advisory panels, shaping the day-to-day understanding and use of those formal structures. Diverse kinds of information are exchanged in the networks, some of it significant, some of it apparently trivial: it is the use of the channel that keeps it open. Arranging for the placement of graduate students and postdocs is a routine but highly significant transaction for the community. The exchange of young physicists establishes long-term rights and obligations between the groups involved. Conversely, the absence of any exchange is a signal to notice.

I asked one group leader at SLAC to tell me where his postdocs were from. He began to list them for me, and then seemed quite startled to realize that most of them were from Europe—something that other people in the lab had noticed and had been commenting negatively upon. He said that I probably should concentrate on other groups, because his "sample was biased." This group leader had visited European laboratories in the previous year and had begun to strengthen those connections for his group, but meantime had not maintained his previous level of exchange with other American groups. Furthermore, the European postdocs would all be returning to Europe. He would not have the responsibility of establishing them in permanent positions; on the other hand, he was not forming the future network ties that placement in American groups would provide. It is important to note that this leader was in a position to neglect his network for a year or two with impunity: two years earlier, he had won international attention for discoveries made by his group. Funding had been assured for a major new facility he wanted, and he was later asked to become the director of another laboratory (a post he declined).

Another group leader spoke to me about a postdoc for whom he was arranging a position. He said the question was whether to place the man in a university group which already regularly used his group's detector, or in a group which usually did physics elsewhere. The first choice would maintain the leader's existing network; the second choice would extend his network ties. He added that in the search for new postdocs he was writing to groups in Argentina, Germany, and England. Like the other group leader, he was very interested in strengthening his international ties, but unlike him, he was also cautious about neglecting his traditional base. Being one of the two group leaders at SLAC who were not American, he was perhaps particularly sensitive to the risks and benefits of international connections.

This traffic in students and postdocs strongly resembles the exchange of women between groups through marriage, which serves as a force in, and source of, some kinship networks. Roger Keesing has argued that:

> in [tribal] societies marriage is characteristically a *contract between corporate groups*. Such a contract entails the transfer of a member of one corporation to residence with the other, and hence the loss of work services . . . Moreover, the corporate group that loses its member in marriage also gives up rights to the children of the marriage. It is best to think of this as a transfer of rights, over work services, reproductive powers, and so on. And since marriage is a contract, a transfer of these rights is usually balanced by material and symbolic response.[1]

Similarly, students and postdocs move on to new positions, usually negotiated by their elders, where, in all likelihood, they will spend the rest of their careers. The fruits of their subsequent labor (in the form of discoveries, equipment, funding, and influence) will not belong to their home groups, although these accomplishments, as well as any failures, will reflect upon their teachers. In exchange for the students' productivity, the home group will receive good will from the new group. This may take the form of other postdocs and students, as well as early access to information about the group's research plans. The extent of exchange depends on the relative status of the groups within their network. At any rate, the exchange of students and postdocs reinforces the tie between them.

Two research groups at SLAC have very strong ties to MIT; a third group works very closely with physicists at Johns Hopkins, Carnegie-Mellon, and Iowa State. Groups from the University of California campuses, Cornell, and CERN often work at SLAC. In December 1977, when SLAC was under pressure to show that it was indeed open to outside experimentalists, the administration pointed out that physicists from sixty-five institutions had done experiments at SLAC, sixteen of which were outside the United States, including three each in Great Britain, Germany, and Japan (KEK, Tsukuba, and Tohoku in Sendai), two each in Canada and Israel, and one each in France and Italy, in addition to CERN. The forty-nine institutions in the United States included American University; Ames Laboratory (NASA); Brookhaven; Caltech; University of California (UC) Berkeley, Davis, Irvine, Los Angeles, Riverside, San Diego, Santa Barbara, and Santa Cruz; Colorado; Columbia; Cornell; Duke; Florida State; Hawaii; Harvard; Illinois; Indiana; Iowa State; Johns Hopkins; Maryland; MIT; University of Massachusetts at Amherst; Michigan; Michigan State; the National Science Foundation; Naval Post-Graduate School; State University of New York (SUNY) at Albany and Binghamton; North Carolina; Northeastern; Notre Dame; Oak Ridge; Pennsylvania; Princeton; Purdue; Rochester; Stanford; Tennessee; Tufts; Utah; Virginia; Washington; Wisconsin; Vanderbilt; and Yale. This is a highly inclusive list; active SLAC experimentalist network ties would extend to about one half of these institutions.[2]

The various networks which intersect at SLAC could all be plotted on this list. Fermilab and Brookhaven National Laboratory (BNL) are the other two centers of networks operating in the United States.[3] Whereas in Japan the networks are organized around university departments of physics, in the United States it is these three laboratories that are the foci of the particle physics community. Each laboratory has a program advisory committee (PAC) that reviews applications for experiments, determines which of the proposals are to be accepted, and establishes how long the experiment should take (how much beamtime is to be allotted). Government funding for research in high energy physics in the United States is granted to the laboratories, not to specific projects: research groups seeking funding need only apply to the lab PAC; they do not deal with the funding agencies directly, so there is no individual peer review. PACs are composed of representatives from the lab, research group leaders from the other two labs, and prominent researchers from university departments. The program advisory committees are forums for negotiation among the networks. Becoming a member of one of the three program advisory committees is a sign of great personal power. Membership is subtly arranged such that each laboratory favors its own networks, in spite of the national representation on the committees.

Ranking of Institutions

A force operating both within and between networks is the stable ranking of institutions. There are eight major national and international research laboratories for particle physics in the world. Two are in western Europe (DESY in Germany and CERN in Switzerland); three are in the United States (SLAC at Stanford, Fermilab near Chicago, and Brookhaven on Long Island); three are in the Soviet Union (Serpukhov, Dubna, and Novosibirsk). All these laboratories are eminent because of the power and quality of their accelerators and the amount of funding they receive. In the first rank now are SLAC, Fermilab, CERN, and DESY, then Brookhaven, Dubna, and Serpukhov, followed by the rest.

Excluded from this list are certain other laboratories, such as Lawrence-Livermore Laboratory, Argonne, Oak Ridge, and Los Alamos, where major research in particle physics was once conducted. Now they are rarely engaged in the most important experiments in the field; there are a few exceptions, seen as such. At these labs, unlike those mentioned earlier, research is also conducted in several other fields besides particle physics, often with equipment originally developed by particle physicists.

The most inclusive listing of top-ranked research centers in the United States would contain the major laboratories (SLAC, Fermilab, and Brookhaven) and a few of the lesser laboratories (Lawrence Berkeley Laboratory and Los Alamos), certain university departments (such as Caltech, UC Berkeley, MIT, Yale, Stanford, Harvard, Cornell, UCLA, and Johns Hopkins), and perhaps ten other lesser schools (such as Wisconsin and Michigan). In Japan, the list would include the Hongo campus of the University of Tokyo, Kyoto, Osaka, Nagoya, and Sendai Universities, KEK, the Institute for Nuclear Study (INS) in Tokyo, and two or three other schools, such as Waseda and Tokyo Metropolitan Universities.

The particle physicists unhesitatingly rank national research communities. For example, American experimental particle physicists consider that the best work is done by Americans, then Germans, English, French, and Soviets (in that order), with the Japanese and Italians about equal. The Japanese are dedicated to moving KEK and their national reputation in experimental work to the first rank. The Americans do not even seem to be aware of this ambition. No American physicist I asked has any clear idea about how such an ambition could be realized. They seem to assume that such a change in relative rank has never been known, forgetting the relatively recent rise of the American and Soviet communities, vis-à-vis Europe.[4]

The only path to the top widely acknowledged by the Americans is the rise of individuals from less prestigious to more prestigious institutions: as some say, "Knowledge trickles down and students percolate up." They are certain that the best people end up in the best places and that low rank means low merit.

This finely structured hierarchy ranks every element of the community. The hierarchy of institutions is presumed to be the outward, visible manifestation of degrees of merit among the physicists, and of degrees of technical sophistication in their detectors and accelerators. The human and technical components combine to determine the rank of laboratories and university physics departments.

Experiment and Theory

Besides rank, at least four other crucial distinctions are maintained in the particle physics community. One of the most fundamental is that between experimentalists and theorists. In principle, theorists and experimentalists at laboratories must work closely together, but they usually exhibit a strong wariness of each other. Their career patterns are different, and they spend their days as physicists differently. Theorists tend to have more status at a younger age; they work in short-lived collaborations with one or two other theorists; they "do physics" at blackboards in their offices. Experimentalists work in long-lived teams from several months' to many years' duration; they "do physics" with their very large and sophisticated machines. Besides physicists, experimental teams include engineers, technicians, machinists, administrative assistants, data analysts, and clerical staff.

Among both theorists and experimentalists there exist many important subdivisions. Phenomenologists, theorists who work at finding the best fit between data and existing theories, may occasionally suggest experiments. A small group called mathematical physicists concentrate on linking developments in mathematics with ideas emerging in particle physics. Some of them may be involved in measurement theory. The majority of theorists, who consider themselves rather superior to the phenomenologists and the mathematical physicists, develop the new models of particle physics.

The most obvious distinction among experimentalists is between groups which address different issues (photoproduction; muon, kaon, neutrino production; leptonic interactions) and use different research apparatus (bubble chambers, spectrometers, colliding beams, cosmic rays). There are also other significant differences between the groups at each laboratory. Group longevity and size, personality of the leader, and the group's access to privileges within the lab, such as funding, computer time, and accelerator beamtime, all vary widely.

Anthropologists have studied many cultures that differentiate themselves into two or more interdependent parts, or moieties, each with specific and binding formal obligations in the culture. The moieties maintain relations by carefully defined rights and duties of intermarriage. They also often practice ritual avoidance of each other in certain fixed contexts. The behavior of theorists and experimentalists in particle physics follows this model quite closely.[5]

Experimentalists and theorists, for example, learn to display a studied disregard for each other's judgment. For example, one experimentalist said to me that "theorists believe anything if it is on graph paper." On the other hand, theorists thought it appropriate that I was concentrating my study on experimentalists: they saw the experimentalists as a rather predictable lot. They assumed that, as an anthropologist, I viewed the experimentalists as a "primitive tribe," which confirmed many of their own assumptions. While having lunch with an experimentalist and theorist at Fermilab, I asked about this studied disregard. The experimentalist said that if their results contradict current theories, experimentalists among themselves presume that something is wrong with their experiment. The theorist said that under the same circumstances theorists assume that something is wrong with their theories. But they agreed that when they are in each other's presence almost everyone acts as if the reverse were the case.

I saw an eloquent example of theorist-experimentalist interaction when a group of experimentalists at SLAC made an important discovery that directly challenged the widely accepted work of an eminent theorist. Very soon thereafter, this theorist came to SLAC to give a theory seminar. Usually at these seminars people sit widely dispersed, except for a few who cluster in the front rows near the speaker. On this occasion, members of the experimental group began to enter the auditorium and sit together in the smoking section, at the rear of the room. They completely occupied three rows, leaving an aisle seat for their leader, who arrived last. During the hour that the theorist spoke, explaining why his theories were not in fact threatened by this new little bit of data, he glanced at the experimentalists only a few times. SLAC theorists interrupted the speaker many times to ask questions. The experimentalist group leader smoked continuously, and no one in his group spoke. At the end of the talk, the group left silently, *en bloc*. The group later won the Nobel Prize for their discovery.

While doing this study, I met the theorist whom I later married. We had been introduced by an experimentalist research group leader who had become a very supportive sponsor of me and my work. This senior experimentalist encouraged our friendship with the same subtlety and persistence I had seen him use in orchestrating projects within his group, the lab, and program advisory committees. Another senior experimentalist, who had been an invaluable contributor to my understanding of both particle physics and particle physicists, told me that he had heard I was becoming close with "a theorist" and asked if it were true. I shrugged and smiled, but he continued to make inquiries. Finally, he said, "Sharon, how could you do that? I thought you had come to understand the difference between us [experimentalists and theorists]!" Both the avoidance and denigration of theorists and the careful orchestration of linkages to them are a significant part of a senior experimentalist's repertoire, and conversely for senior theorists. Younger physicists learn the rituals of avoidance first, before they learn how to forge and maintain appropriate links. The links are often in the form of personal friendships between physicists at different laboratories, sometimes through spouses who have become acquainted at laboratory social functions. It is not uncommon for senior experimentalists' daughters to be married to theorists, and vice versa—not only in the United States, but also in Japan, Europe, and the Soviet Union; but I know of only one experimentalist-theorist marriage (they are Americans and the wife is an experimentalist).[6] I was startled to find that many of the younger experimentalists had so few ties to theorists that they did not hear about it when I married one. When they heard of my marriage, the senior experimentalists began to ask me what my husband's position was on certain issues at his laboratory and arranged through me to discuss physics with him.

Research Groups

I have already suggested that strict boundaries are maintained between experimental research groups within a lab. The younger physicists are the most insular. Several said that they knew few physicists at their laboratory outside their group, and that they knew little about the other groups. Those who had worked at CERN in Switzerland remarked that they had met many more physicists in other groups there. Several commented on the fact that CERN had a pleasant gathering place where "cookies and good coffee" were available, and that is where they had met others. Even though little physics was discussed in these encounters, they had felt better informed about the laboratory there than at SLAC. One mentioned that he met with the SLAC director upon his return to suggest establishing such a place. He said the director listened carefully, but then did nothing. He believes the insularity of groups at SLAC is extreme, and is perhaps encouraged in order to reinforce the authority of the "old boys' club." Two young European physicists visiting SLAC for one year asked me how the groups interacted. After several months, they said, they had been able to detect little communication.

The various research groups do not mingle in the research yard. Each group has its own coffee urn; when it malfunctions, they do not "borrow" coffee from a neighboring group. I was the only person I ever saw moving freely and unannounced through all parts of the research yard. After I had been at the lab about a year, people began asking me occasionally how things were going in the other buildings. The only times they visited other people's detectors were for the parties: the birth of LASS, the demise of the eighty-two-inch chamber, and the arrival of the new particles at SPEAR. Even then, they would stay at the edges of the gatherings.

An experimentalist at Fermilab said he thought that group borders were strong everywhere, but that perhaps they were more noticeable at SLAC because the groups there were so long-lived. At every lab I visited, I saw that each group had a stockpile of currently unused equipment. I also learned that each group builds its own computer software package entirely "from scratch." When they need a scintillation counter, a magnet, or software, they do not ask another group if it has one available in storage which can be borrowed. When asked why the groups did not share even the most rudimentary equipment or software, one experimentalist replied that "we are just constitutionally disinclined." Some groups do try to "raid" good technicians from other groups, but they are rarely successful. Groups are very careful to manage the news that leaves their group, and to appear not to need any news from outside. When data from one group's major set of experiments were looking increasingly nondescript to them, some of the members became very anxious and the group became much more secretive. I learned that a few of the members began looking for other positions, hoping to get established elsewhere before news of the group's decline became widespread.

There are people at every laboratory who provide communication across these borders. At SLAC there is one physicist who has a joint appointment in two groups that use the same facilities. However, his status is fixed, and he probably will find advancement to the top difficult because divided commitments and loyalties put him at a disadvantage in a very competitive atmosphere. A few experimentalists provide "liaison" for the visiting user groups, arranging for the support services they need to conduct their experiments. This position also has low status and little opportunity for promotion. Administrative assistants to the group leaders know one another well, and are a means of disseminating and gathering information informally, but they too have low status. Phenomenologists usually have much greater contact with experimentalists than other theorists do; and with a few notable exceptions, phenomenologists have lower status than other theorists. At every laboratory I visited, a high proportion of these intermediary positions were held by women. In highly structured occupational environments, in general, women are typically assigned to tasks at the borderline.

Women and Men

The sharp division in labor among men and women physicists is not distinctive to SLAC or the United States, but it is more acute in particle physics than in almost all other scientific disciplines. For example, even though many physicists in the Soviet Union are women, few are in experimental particle physics, and I heard of none in theory groups. One Soviet male theorist told me that he did not expect his daughters to become theorists, because women's minds were not suited to the work. In the Soviet Union, each work group has tasks to perform for the community beyond their job responsibilities. In one solid-state physics lab in Leningrad all employees were expected to pack vegetables one day a week for distribution through the city. In practice, though, among the physicists only the women actually performed this task.

In Japan, in both laboratories and physics departments of universities, women perform the clerical work. Work which is done in the United States by administrative assistants is all done by group leaders in Japan. The department secretary often shares the same office as the chairman. As in the Soviet Union, one of the clerical workers' jobs is to prepare tea. In Japan the department secretary also acts as hostess when the chairman has guests.

I saw very few women technicians in Japan, although most libraries were run (if not officially directed) by women. Part-time work (*arubaito*—a Japanese word taken from German) is a common arrangement. Scanning is done by part-time workers. There is a problem at KEK: I was told that, since KEK is in a rural area, there were no local people well educated enough to be trained for scanning except the wives of physicists, "who, of course, are inappropriate." Physicists are concerned because they must ask students to do this job.

The number of women theorists in Japan is proportionately the same as in the United States—about 3 percent. When I asked why there were no women experimentalists, I was told by men that it would be inappropriate because experimentalists must often work at night, which until 1986 was illegal for women. The professional life of women physicists is complicated by the fact that much physics discussion takes place when the group goes to dinner. In Japan all work groups socialize very frequently, sometimes several times a week. I never saw a woman in these groups. It is considered inappropriate for women theorists to join men theorists in a public social context, so the women theorists miss valuable discussions.

Among the few women worldwide who are particle physicists, there appear to be some similarities. The Europeans and Asians seem to be from upper-class, privileged families; many of the Americans are immigrants or children of immigrants. Two possible explanations come to mind. First, in societies with very strong class systems, upper-class status may compensate for the low status of being a woman, particularly if her colleagues are of lesser social class background. Second, for a few generations immigrant families may be so concerned with survival and raising their status that they allow their daughters to participate in this effort to their full abilities in spite of their traditional gender expectations. Even so, everywhere in the particle physics community, gender is a difference that makes a difference. Traditional sex roles in the broader society continue to determine the division of labor in research as well as apportioning power in decision making and in forming models of success and failure: women remain at the margins.[7]

Talking and Writing

The last of the distinctions to be discussed in this chapter is that between restricted information, which is spoken, and public information, which is written. What is accomplished among the physicists by talking? Primarily they are evaluating their peers and their work, persuading those same peers to support their own work, managing the distribution of news, arranging positions for novices. These agonistic evaluations of other physicists and this process of enticing others to corroborate one's own data stand in marked contrast to the image of objective, neutral decision making in science that one finds in the influential analyses of Max Weber and Robert Merton.[8] Oral communication is fundamental to the operation of the particle physics community and successful senior physicists are masters of the form.

In conversation physicists are often engaged in calibrating their evaluation of a piece of work with others' opinions. In any given particle physics experiment "error bars" are calculated; these error bars refer to the proportion of data generated in an experiment that cannot be associated definitely with the phenomenon under study. There are various ways of calculating error bars (such as Monte Carlo analysis); it requires very sophisticated knowledge of a detector to do it properly; and, furthermore, the procedures are not firmly established. So people are heard to say, "Oh, whenever I see their work, I multiply their error bars by three or four and see if the results are still interesting." It is this factor of three, four, or, in the exceptional case, five which is orally debated. I have never heard a group discuss what they think their own error bar factor is, as seen by outsiders. It is said that if one has "total, absolute trust" in an experimentalist, one should nevertheless multiply his error bars by at least a factor of three.

Physicists also discuss whether to pay any attention at all to colleagues' work. Several people within and outside SLAC referred to two groups there as dead wood, which would never be supported if their leaders did not belong to the SLAC old boys' club. Commenting on two groups which use the same facility at SLAC, a physicist said that one group—currently very powerful and successful—only "does hardware," whereas the other "does physics."

Talking is not only for evaluation. Talking is the way to get something done, to win extra computer or beamtime for one's group, to acquire a good postdoc, or to persuade other physicists of the significance of one's own research results. Persuasion is a critical skill. I have been told of interesting experiments that no one has bothered to corroborate because "the group didn't go out talking about their work. If they don't believe in their work, why should we?"

It is possible, though, to argue forcefully on behalf of one's work in ways that the physics community will regard as inappropriate. A SLAC postdoc told me that one night he was monitoring an experiment with one of his group's collaborators, a professor from a nearby university. Very interesting data began to emerge, and with great excitement the professor exclaimed to the postdoc, "We're seeing the first exotic meson!" He believed that everyone would now spend years looking for the rest of this family of particles—"We're the beginning of a factory!" The professor said he knew he would now become leader of a new group, and that he wanted the postdoc to be the first member. The postdoc said he just kept quiet, because he felt very skeptical of the new idea. During the next few days, the professor called together a regional meeting of the whole collaboration of the in-house group plus users, and made a very dramatic presentation, announcing: "Gentlemen, I [the postdoc noticed he did not say 'we'] have seen exotic mesons." When the professor then showed his data, it was felt to be outrageously inadequate. His mesons could well have been just noise. His audience became very angry, and eventually the professor lost his promotion. The postdoc said he learned that even if you think you have found something, you should hold on to your data and make *very* sure before you make claims, even at the risk of losing priority.

It is also possible to criticize another's work inappropriately. I learned that one theorist from an American university had acquired among several theorists at international laboratories a reputation for asking everyone in detail about their current work. At the end of the questions he would say that he had been working on the same topic with the same general conclusions and suggested that they write up their work together. Since short-lived collaborations, very common among theorists, are often initiated in this way people would usually agree to this suggestion. Then they would discover that this particular university theorist had very little to contribute. They would complete the ostensibly coauthored article but resolve to avoid him in the future. When the laboratory theorists, meeting at conferences and during visits to one another's laboratories, began to compare their experiences with this one man, this informal exchange sufficed to reduce his easy access to many people and the number of his publications began to drop. When this man was being considered for a promotion in his department, one laboratory theorist volunteered this story to another physicist in the department. The lab theorist was asked by a department representative to provide the names of corroborators, but when pressed the other lab theorists he named declined to comment, expressing dismay that this confidential information had been passed to "outsiders."

A physicist at SLAC was avoided for weeks in the cafeteria because he had communicated with reporters from a local newspaper about his work before publishing his results in a physics journal, and the other physicists believed he himself had initiated the contact. A very significant time elapsed before he could persuade other groups to corroborate his data; eventually its importance was established. When another group at SLAC found data that eventually led to the Nobel Prize, the primary journal in the field nearly refused to print the report because during the weekend that the discovery was made, a student reporter had gathered some information about it and his story had been printed in the university newspaper. A group member explained to me that it was eventually settled by the group's assuring the journal editors that they had not sought out the student reporter and that he had made his story out of casual remarks uttered in the excitement of the discovery.

In Japan the particle physicists who want to increase funding for experiments at laboratories outside of Japan have tried to impress Monbusho (the Ministry of Education, which makes all decisions about funding of basic research) with their credentials, but here too the constraints on legitimate self-promotion are rigid. One physicist showed me the telegram he had arranged for his foreign collaborator to send to Monbusho; the telegram strongly emphasized the significance of the Japanese contribution to a recent discovery the collaboration had made. He stressed that the telegram had been designed to impress Monbusho. Others told me that this initiative was strongly condemned by his colleagues and that his funding would probably be cut.

The two incidents at SLAC and the Japanese telegram were seen as efforts on the part of physicists to influence the particle physics community by influencing forces outside it. Transgressing the boundary of that community in such ways is completely unacceptable. I have seen no such efforts succeed, even when they could have been seen as sanctioned by another value in the community: the right to claim priority in scientific discoveries. The "transgressors" were surprised by the response to their actions; in following one value they had violated another. The obligation to observe boundaries and not to talk to outsiders clearly takes precedence over the right to claim priority.

Access to this world of oral communication is quite limited. In a community with easy access to widely disseminated written information, keeping crucial information accessible only in oral form is an impressively effective means of maintaining its boundaries. There are many precedents in many cultures for the use of oral communication among the literate to restrict knowledge to the initiated and the elite. Maimonides cautioned his students:

> It is not permitted to divulge these matters to all people. For the only thing it is permitted to divulge to all people are the texts of the books. You already know that even the legalistic science of law was not put down in writing in the olden times because of the precept, which is widely known in the nation: *Words that I have communicated to you orally, you are not allowed to put down in writing*.[9]

Protection of oral communication encourages the development of a closed community. In physics it is consistent with the group's image of itself as a meritocracy: only an informed, worthy member of the community will know what is to be said and what is to be written.

Good experimentalists do, write, and talk physics, but they rarely read physics. Important results are usually written up quickly and are available as "preprints" within a few weeks of discovery or innovation in detector or software design; preprints appear as journal articles within months. Particle physics changes so rapidly that waiting to learn of interesting data, detector innovations, or new theoretical developments until they appear in the journals is regarded as exceedingly unwise. What is being talked about is the current, more advanced knowledge; what has been written is considered established, uncontested, and hence uninteresting.

Particle physicists do scan preprints in order to know who is writing about what.[10] If something catches their interest, they will phone or waylay the author to try to elicit, preferably face to face, what they want to know; it is assumed that the whole story is rarely written. They want to know what is going on and who is working on what before it is written; it is also thought to be important to know who is planning on working on what, at which lab. They also want to know who has the inside track on getting beamtime at which labs. (There is always unscheduled beamtime to be allotted to qualified groups if they have a very promising new proposal.) They assume that all this can only be discovered by talking to people.

A physicist joked to me that this talk among physicists was like "photons being exchanged among interacting particles." Exchanging judgments about one's peers, persuading colleagues to support one's work, managing news, being a competent performer of combative, tendentious jokes (preferably using technical language from particle physics to describe human behavior), and being an informed gossip are crucial skills for a successful particle physicist.

Only particle physicists have the capacity to engage in this talk and only they (and their staffs, spouses, and an occasional anthropologist) have access to it. Knowing the stories and performing in the appropriate style is an unmistakable sign of being a real particle physicist, of knowing particle physics, and of knowing how to make knowledge about particle physics. It is through this talk that physicists reach agreement about who is to be a particle physicist, what is a good research instrument, what will count as a fact, and who should be allowed to try to make new machines and facts. Everyone's future depends upon the collective outcome.

This talk is judgmental: real sanctions and rewards are visited upon its objects. High energy physicists need good reputations to stay in the community; loss of a good name means isolation and even expulsion. This is a small face-to-face community, even though an international one—a dispersed village—and there is no other community one can turn to for support. Very few who incur disapproval escape the sanctions. To be beyond criticism one must either be extremely powerful, like a Nobel Prize winner, or quite powerless, like a graduate student. Of course, one can leave the community, become an astrophysicist or biophysicist; high energy physicists consider such defectors the "leavings" of their field, with all the associations with contamination that suggests: "no one in high energy physics talks to them." Physics and its culture is produced and reproduced through talk which is storytelling, talk which is judgmental, talk which punishes and rewards. In short, by gossip.[11]

In conclusion, talk accomplishes diverse tasks for physicists: it creates, defines, and maintains the boundaries of this dispersed but close-knit community; it is a device for establishing, expressing, and manipulating relationships in networks; it determines the fluctuating reputations of physicists, data, detectors, and ideas; it articulates and affirms the shared moral code about the proper way to conduct scientific inquiry. Acquiring the capacity to gossip and to gain access to gossip about physicists, data, detectors, and ideas is the final and necessary stage in the training of a high energy physicist. Losing access to that gossip as punishment for violating certain moral codes effectively prevents the physicist from practicing physics.

The question remains: why don't they write what they say? If gossip is a means of producing physics, physicists, and their culture, then written materials, articles and preprints, are the commodities the physicists produce in their turn. Articles represent the consensus, the "facts," data with the noise removed. The authors of these written accounts own the information in the account. Any subsequent users of that new information must pay royalties to the authors in the form of homage or credit, thereby increasing the accumulating reputations of the authors. In talk physicists rarely give credit to others. Scientific writing keeps track of the results of these debates. It is a record-keeping device, a spare ledger of credits and debits. Citations are a trace of something happening elsewhere, as a bubble chamber photograph presumably records traces of events in subatomic nature.

In this chapter I have argued that several sets of distinctions are articulated throughout the particle physics community. Experimentalists and theorists are sharply differentiated, borders between groups are carefully observed, the occupations of men and women are highly segregated. Networks of exchange link otherwise autonomous units at every level of social organization. The primary commodities exchanged are students, postdoctoral research associates, and "gossip" (oral information about detectors, proposals, data, organization of groups and labs, and the location and professional genealogies of individuals). The boundaries of the networks as a whole are closed, marking off the outsiders. Finally, oral communication is of great importance to the community and serves a role clearly distinct from that of written communication. These people know a great deal about one another's work and lives; gossip about both is widely exchanged. They meet frequently at large and small conferences and committees; they have free and easy access to each other by telephone; and their work is disseminated to one another very quickly. Particle physicists note great disparities within their community in access to this oral communication, but compared to other academic groups the emphasis on oral communication is marked at every level of their hierarchy. The boundaries of the community as a whole are negotiated with great circumspection.

Objectivity and Traces of Reason

In this chapter especially, my attention has been directed away from physics, from objects of knowledge, and toward physicists, subjects of knowledge. The practice of science depends on reducing all questions to problems involving a finite set of variables, by means of conventions that exclude certain aspects of an event from what needs to be explained. This exclusionary habit of mind is part of how scientists see themselves as scientists, people particularly suited to understanding nature.

My conversations with theoretical and experimental physicists suggest that they characteristically see each human being as a composite of rational and irrational elements, of order and disorder. They also believe that the proportion of rationality is randomly distributed among humans, and that curves of random distribution have certain predictable characteristics. For instance, there will always be a small proportion of humanity with an exceptionally high rationality quotient; these are the people who are suited to become scientists. Not all of them may in fact end up as scientists, but all good scientists come from this "gene pool." On this model, scientists are not made but born—and then revealed as scientists through a series of exclusions made in the course of childhood and early scientific training.

In nature too, as viewed by the physicists, under the appearance of randomness and instability lie regularities, immutable mathematical laws. The classification of phenomena reductively, according to a limited set of attributes, is seen as a necessary first step in discovering this inherent order. Anything that eludes reduction to the chosen attributes, whether in nature or in human nature, is avoided, rejected, or considered marginal or comic. Underlying regularity is a necessary assumption if nature is to be knowable—and if scientists are to be identifiable. Progress, in science and in the education of scientists, depends upon it, since regularities are to be revealed, not constructed.

In nature the same forces and the same elements exist in the same proportions for all time. The periodic table of the elements and "SU3," the classification system of fundamental particles, both display immutables and their relations. All apparent change is merely the redistribution of basic components; the fundamental particles appear to be transformed into other particles, but these transitions or "decay modes" are only patterns of recombination. The scientific template for cause and effect seems to suggest an inherent temporality; but since effects are seen as inherent in causes—just as a particle's decay mode is inherent in its description—all change is reduced to recombination. Like budding scientists in school, nature awaits discovery. Nature does not acquire order, but is intrinsically orderly; scientists do not acquire rationality, they are rational by definition.

By contrast, whatever is inappropriate or imperfect is cut away, to fall outside the system. In interactions between particles, instabilities are short-lived and rapidly move toward stable configurations. Scientists may be in error—their errors may even be accepted for a time—but error is unstable and will be progressively eliminated.

The language of physicists is rich in negative images of change—deviance, decay, annihilation, fluctuation, instabilities in the beam—and centered on positive images of stability. The "simple initial state" has to be established as a precondition for any experiment, often with great difficulty. In particle physics, this desired "ground state" requires, first, months and even years of meticulous detector design and construction, followed by elaborate computer analysis of the statistical errors and systematic errors that are inherent in the design and operation of the detector, the accelerator that feeds it, and the experiment itself. The running of the experiment is a painstaking business, with continual calibration of the accelerator beam, the detector itself, and the data collection process—all in pursuit of stability in the "ground state," the background against which "significant events" are to stand out.

In these ways, the experimentalists begin their task of deciding what part of their data can be considered valid and what must be "cut," discarded as "noise." Beyond the establishing and maintaining of the simple initial state comes the analysis of data. Interpretation of traces is quick and straightforward only in rare cases. It takes human, linear time to pull the physics out of the morass of data: usually a year or more, after which the results are scrutinized by the community and accepted only when confirmed by data from other experiments. For other experimentalists to be interested in corroborating the results of an experiment, they must first be persuaded of their significance. This depends in part on their confidence that a precisely specified ground state was indeed generated as background to the reported results.

This model for good physics is very like the model for good physicists, and for a good environment for physics. Whether in training postdocs or in running a laboratory, "predictable, smooth behavior" must be established before instabilities are welcomed. Only against the background of a fully specified ground state can new ideas or institutional change be accepted as productive. It is this concern for the "coherent ground state" that leads a senior physicist to demand that postdocs be meticulous and thorough, hard and willing workers with the patience to do a very long and tedious job carefully. The same concern is invoked by directors of laboratories when resisting change in their organizations.

In spite of all efforts to maintain stability as a prerequisite for truth, uncertainty and error remain present and acknowledged in scientific work. According to the physicists, this is because as human beings they inevitably are subject to some admixture of irrationality, even if in minor amounts compared to the rest of the population. Pure objectivity is tacitly recognized as impossible; but error can be estimated and minimized. The means is peer review, or collective surveillance; the final degree of order comes from human institutions.

CHAPTER · 5

Buying Time and Taking Space: Negotiations, Collaboration, and Change

The members of the particle physics community are firmly committed to the international, supracultural image of science.[1] Particle physicists from anywhere in the world are fond of remarking that they have more in common with each other than with their next-door neighbors. All of these physicists consider themselves members of an intellectual elite, perhaps *the* intellectual elite, because they believe particle physics works alone at the frontiers of human knowledge. Shared traditions in training and in distinctions of the kind discussed in Chapter 4 serve as stabilizing elements in the international community. It is in the context of this relatively constant structure or "ground state" of predictable smooth behavior that problems in the community can be defined and sometimes resolved.

The overriding issue confronting both Americans and Japanese is how to incorporate new facilities for doing physics, such as PEP, the new collider at SLAC, and KEK, while maintaining the stable structure of their world. In other words, can these communities reproduce themselves in new contexts? The experimental particle physicists in both countries debate the model of a good laboratory in terms of "the best environment for good physics." In both cases the pivotal issue is the relationship between in-house groups and "users." In the United States, SLAC is considered the American laboratory most strongly controlled by inside, permanent research groups.

In-house and User Groups at SLAC

Most of the eight SLAC group leaders have been at the lab since its inception. Relations among the groups are highly, though informally, structured. "In-house" groups are ranked; the ranking fluctuates somewhat, but there is a definite status difference between the top four or five and the others. Each of the groups is a long-lived (ten to fifteen years is typical) and close-knit enclave, and the group leaders take an active, round-the-clock role in the operation of experiments and the administration of their group. The groups compete intensely for lab resources, but they also have a strong sense of their collective identity, in distinction to other laboratories and to the visiting user groups from other institutions. As one leader puts it, "SLAC consistently does strong, programmatic physics." Another leader was more specific:

> Our group has a new experiment under way now. I had to get beamtime away from [another SLAC group leader], but the data from [our detector] will be worth it. The data won't be a huge, long-lasting difference, but for a few days it will make a big hubbub. [We SLAC groups] know our detectors: we can take the things apart and put them back together because we know everything about those detectors. A user couldn't do that. The users were very upset about our group dismantling our detector for this experiment; they wouldn't know how to do it. A group can only do physics as good as their knowledge of the detector. When labs are run by users, they do less physics, much less. It is interesting for me to be doing physics with [one group leader] on one side and [another group leader] on the other. I learn. But it's not interesting with most of the users. The users will fade away because we do better physics.

At SLAC the user groups have mounted a large and so far successful attack on what they consider the in-house monopoly on SLAC resources. The result is that the new facility at SLAC is now managed jointly by SLAC people and outside users, theoretically with an equal share of control. The actual practice of this new administrative structure is in an early stage. At a meeting in early 1978 the SLAC program advisory committee (PAC) decided which groups would be allowed to do the first round of experiments at PEP; a collaboration from the midwest was accepted and two SLAC groups were turned down. Outrage, bitterness, and turmoil resulted at SLAC. A common complaint heard that week was, "Why don't people from the midwest work at Fermilab?" This regionalism is revealing; it seems to confirm the long-standing complaint that SLAC is not sufficiently mindful of its responsibilities as a national laboratory. At issue is, as everyone phrases it, "What kind of environment is best for physics?" The SLAC people are convinced that the old regime is best; they compare their experimental results over the last two decades with other laboratories and note that just as SLAC is being forced to abandon its model, to them an obviously successful model, other laboratories—notably CERN and now Fermilab—are visibly moving toward it. At SLAC, resident experimental groups claim that as designers, builders, operators, repairers, and modifiers of the detectors, they are simply much better qualified to know what physics can be accomplished with those detectors—and to accomplish it. They say that they know this is not "fair"; they add that "good science is not made by means of morality or majorities." The users insist on a more democratic approach, which includes more parties to the decision-making process. They claim that more candidates mean more competition, and more competition means a better product, which is to say better physics. I did not hear Japanese physicists using economic imagery in relation to these issues.

An experimentalist who had been a user at SLAC felt that the director at that time had maintained the in-house group dominance by "buying off" the most powerful of the users so that the user groups as a whole remained docile and weak. He added that the former director of Fermilab would have been "hung" if he had acted as the SLAC director did. From the SLAC point of view the director had simply been maintaining a well-established and productive "coherent ground state." Once an important experimentalist, he has been a fierce and powerful presence in a world of men who assiduously cultivate the postures of power.

By mid-February 1978 the SLAC director had begun to counter the users' initial gains. He noted in a memo that the original "Cooperative Basic Agreement" between SLAC and Lawrence Berkeley Laboratory (LBL), a powerful group using SLAC, concerning the management of PEP had stated that the "parties may enter into . . . agreements implementing this agreement as may be necessary." He arranged for a new division to be formed within SLAC with the responsibility for operating PEP; in the execution of that responsibility the new division was to involve LBL only when it was "appropriate." The PEP division was to be directed by a new associate director of SLAC, who would be appointed by the SLAC director and would report directly to him. The assistant director of the PEP division would be appointed by the LBL director with the approval of the SLAC director, and would continue to serve with the consent of the associate director. The present program advisory committee for PEP would be combined with the SLAC PAC into a new Experimental Program Advisory Committee (EPAC), which would have eleven members, jointly appointed by the SLAC and LBL directors. Six were to be high energy physicists who were not employed at SLAC, LBL, or UC Berkeley; three would be chosen from six persons nominated by the joint SLAC-LBL users' organizations. The chair of this EPAC would not be from the SLAC or LBL staffs. It was noted that:

> The EPAC shall be constituted to review experimental proposals under procedures designed to assure equitable access to the entire high energy physics community, while maximizing the physics productivity of the entire SLAC/PEP complex.

Furthermore,

> all decisions to accept or reject a particular proposed experiment shall be made by the director of SLAC or his assigned deputy after receiving the advice of EPAC . . . and after consulting with the LBL director.

This draft was accepted, with certain compromise provisions, by the end of February. SLAC had relinquished some of its autonomy by joining forces with LBL, but that alliance enabled the two labs to regain at least some of the ground lost to the users. Rather than open the lab to "democracy" and the decline in the quality of physics that he assumed would result, the SLAC director moved to reestablish the lab's traditional hierarchical decisionmaking structure. The analogue in Japan would be if KEK and INS—the Institute for Nuclear Study at the University of Tokyo—were to combine forces against university departments of physics by means of an agreement between the two directors. In Japan such an agreement could only be the result of massive negotiations; once made it would not be hard to implement. In the American case, it was a brilliant tactical move; it remained to be seen whether the agreement would achieve the ends the director hoped for.

The regularly scheduled SLAC users' conference was held at the end of March. Usually these meetings were largely ignored by the SLAC administration and group leaders. Under the new agreements the users were entitled to be informed of policy decisions under consideration so that they could offer their "advice." Issues which would ordinarily have been discussed and settled in the director's office were now being raised in the auditorium with an audience of one hundred. This users' conference reminded me of the first televised national conventions of the two main American political parties, in which the parties had not yet comprehended the impact that the media would have on their operations. The director had managed to preserve the lab's hierarchical decision-making structure at the cost of making the process "public."

Representatives from SLAC groups described their detectors and the interesting physics questions that they proposed to explore with them. The director explained the "organization and committee structure after the PEP turn-on." Two competing proposals for a new detector to be built for PEP were heard. (The advocate of the accepted proposal would become leader of a new group, in accordance with a "traditional" career path which had long been inaccessible at SLAC.) In the discussions that followed each of these presentations, the director and several of the senior SLAC physicists actively participated, making barbed jokes about one another's present and past work. The users generally did not participate in this banter, which clearly reflected the established hierarchy between the SLAC groups: the person with more authority always had the last word. This ritual combat in the form of bantering insults displayed both aggression and intimacy among the participants.

From about the time of this meeting people at SLAC began to refer to the groups which had had their experimental proposals for PEP denied as "unemployed." A month later, a successful young SLAC physicist said that SLAC "may be on the way down, as of today." This was in response to the news that the colliding beam facility at a German lab had confirmed the existence of a new particle, the upsilon, first detected at Fermilab; the physicists at SLAC assumed that if their organizational structure had not been undermined they would have identified the new particle. Since then, PEP has been completed and begun operation. An analogous facility (PETRA) in Germany has also been completed. I have learned that many physicists believe that the physics "coming out of" PETRA is much more interesting than that done at PEP; furthermore, it is said that the Americans have fallen behind the Europeans in the last three to five years. The physicists at SLAC are angry and bitter about these developments and trace the decline to the users' revolt at SLAC. SLAC is widely considered to have a special *élan vital* in doing research. The physicists at SLAC agree and that is why they are so offended by an effort to make them revise their customary procedures.

INS and KEK

One Japanese physicist told me that he thought KEK was a "museum piece" before it was built, that it serves primarily as a "monument to the older generation." He pointed out that for reasons of "diplomacy" many people who are not really interested in KEK will be obliged to do an experiment there. He compared KEK to Argonne in the United States, implying that KEK was as obsolete as Argonne, as far as high energy physics is concerned. This man is one of a rather large group of high energy physicists who are based in Japan but who do their physics research abroad at the major laboratories in the world. Many such physicists would have preferred that KEK's funding had been spent on sponsoring Japanese physics research abroad. Some of them concede that KEK could be useful as a testing ground for the design and execution of experiments to be carried out at higher energies at other accelerators abroad, and that this would provide excellent training for young physicists.

Several physicists at KEK urged me to visit another laboratory in Japan, the Institute for Nuclear Study (INS) at Tokyo (a branch of Tokyo University founded in 1953).[2] One man said that he had found "a very good physics atmosphere at INS" when he worked there as a graduate student. He claimed that it was at INS that he learned his "approach to experiments" and his "attitude towards physics." Another physicist, based at a Japanese university, supported INS because he believed that small machines were best for training graduate students, who should in his opinion learn all aspects of experimental work. This man had not worked at INS himself; in fact, he had worked as a postdoc at the small, now dismantled Princeton–University of Pennsylvania accelerator. I later learned that this attitude was one strongly and widely held in the United States by those who had worked on that machine. Many senior staff physicists at SLAC had trained there.

At INS I was introduced to a research group leader whose office was like those of the senior faculty members at university physics departments, including the small couch, low table, and tea. Others in the group joined us for a lively discussion. The INS electron synchrotron was built between 1956 and 1961. The researchers were using a Toshiba 3400 computer, similar to a Sigma 5 at SPEAR.[3] At the time of our meeting INS had submitted various proposals to the government for modifications of and addition to their facilities. They had received budgeting for new computer components; there is competition with KEK over the other proposals, such as a polarized target, a colliding beam facility, and a synchrotron radiation facility. When I asked about the competition with KEK, there was a silence and several sideways glances were exchanged. Finally, the group leader said that many KEK people were formally at INS. I then volunteered information about persisting tensions between a laboratory and a university department in the United States that were based on funding conflicts in the 1950s. The group was very interested in this issue, and asked several questions. When I asked if there were parallels, they pointed out that the decision would not be made for four or five more months.

On another occasion I asked a senior physicist at KEK about the INS-KEK competition. By that time, I seemed to have acquired a reputation for reasonably good Japanese manners; he was clearly startled by my raising this question. After I asked another, more specific question, he responded that there are no other research institutes on the financial scale of KEK; the others will watch KEK and follow its example in methods of seeking funding, and so on. He added that Japanese do not like competition, that Japanese do not like anyone or anything to get too high or too low, that no one should be excluded from the "golden mean." I gathered that he found this attitude simultaneously frustrating, proper, and a matter of only slight interest. His tone of voice was dispassionate, but his shrugged shoulders and lifted eyebrows showed impatience and resignation. He pointed out that there is general expectation that physicists in Japan will avoid intruding upon one another's specialties.[4] The KEK physicists interjected that the people at INS are really very good high energy physicists. When KEK needs work done that does not represent any specific particle physics specialty, they usually call upon a high energy physicist from INS.

I later learned from Japanese physicists in the United States that a major 2.5 GeV synchrotron radiation facility was to be built at KEK. (A much smaller 300 MeV was completed at INS.) This facility, called the Photon Factory, was funded in April 1978 and the first photon beam was expected by late 1981.[5] This decision, according to the Japanese visitors, continued to be "a big problem" for the Japanese physics community. Synchrotron radiation facilities are used as research equipment for materials science; they make use of radiation in the form of photons, a by-product from the acceleration process. The photons are used like an extremely precise X-ray device, to determine the internal structure of organic and inorganic substances. The Photon Factory will be used in part to develop techniques for fabricating higher-density integrated circuits.

The synchrotron radiation facility at SLAC (the Stanford Synchrotron Radiation Laboratory, or SSRL) is appended to SPEAR, the colliding beam facility. Unlike the rest of the laboratory, which is funded by the Department of Energy (DOE), SSRL is funded by the National Science Foundation (NSF); the difference in funding is a signal that very different kinds of science are done at SSRL and at the rest of SLAC. Nevertheless, the SSRL funding helped in the development of SPEAR, just as the expanded SSRL will help in the funding of PEP.[6] At SLAC, the funding of SSRL did not represent a challenge to another laboratory or department in the particle physics community, as the Photon Factory does at KEK. In neither case is synchrotron radiation of significant research interest to particle physicists, although the construction of such facilities is a challenge for physics instrumentation groups. The significance of the Photon Factory for the Japanese physics community is that it established that KEK was to be funded more generously than other physics research institutes in Japan, but not on its own terms: the high energy physicists would have preferred that the money be used to upgrade their accelerator, rather than to build the Photon Factory.

The emerging prominence of KEK, which is affiliated with the new Tsukuba University, over INS, which is affiliated with the dominant Tokyo University, also represents a challenge to the influence of Tokyo University in national government funding policies. Some of the senior scientists at KEK spoke quite intensely about their sense of the need to dislodge the power of Tokyo University; this goal seemed to account for at least part of their own decisions to work at KEK in Tsukuba.

There is a comparable problematic network relationship between SLAC and Stanford. In the late 1950s and early 1960s conflicts developed between certain members of the Stanford physics department who advocated the establishment of SLAC, and others who believed that funding such a large project surely would divert support from other physics research. Furthermore, there was concern that placing the new laboratory at Stanford might eclipse the physics department's other projects. The original reasons for the conflict have receded for the most part; but its traces remain in habits of distance between SLAC and the Stanford physics department. SLAC is not a part of the department, although a few physicists at SLAC have faculty appointments in the department; other faculty appointments at SLAC are Stanford University positions not associated with the department. There is a tacit understanding that SLAC will concentrate on postdoctoral education and avoid acquiring many graduate students. Stanford high energy physicists tend to work as users at Brookhaven or Fermilab. When I asked about this delicate relationship, SLAC physicists usually said there was no longer any real conflict, but added that this awkward problem would end completely only when the participants in the original dispute (who are now in their midsixties) are no longer living. The Japanese physicists with whom I talked were especially interested in these conflicts; they thought there were many parallels to the tension between KEK and INS and between KEK and the universities.

Tsukuba: A New Model for Physics Education

One physicist who had done much of his research abroad spoke of this tension in terms of Japan's "next crisis." He argued that in the future Japan must rely on intelligence, talent, and innovation in order to remain independent. His point is one I heard often in

Japan after the "oil shock" of 1975: Japan cannot achieve self-sufficiency in raw materials, energy, and foodstuffs; hence, Japan must produce high-quality items for export to earn the currency to buy materials that cannot be produced in Japan.

He went on to say that the present system of higher education is too tradition-bound to help Japan into the future. In the opinion of this physicist and others I spoke with at KEK, Tokyo University and its former students in the government bureaucracy control the educational system in Japan. They believe that Tsukuba University represents an attempt to create the new education Japan needs. I was told that Tsukuba admits students on the basis of recommendations and not exams, a nearly revolutionary innovation for Japan. But Tsukuba seems to be having difficulty recruiting good faculty and good students. One physicist said that it is important to recruit not only good people, but people who are not "confined by their need for security." He believes that preparation for the university examinations all through the school year leaves very few students who are adventurous or who are exceptional in any one area (as opposed to those who demonstrate a high average level of competence in many fields). It is his opinion that Tsukuba, too, ought to try for excellence in a few fields.

Several physicists mentioned that the structure of Tsukuba was modeled on the University of California (UC) system, and in particular on UC San Diego. (UCSD began as an outgrowth of the Scripps Institute of Oceanography and continues to have a strong emphasis on the sciences. It also has had an innovative administrative structure and curriculum.) One man is credited with establishing Tsukuba as a university and a science research center. He is said to have close friendships with various officials in the UC system. Informal discussions at Tsukuba often come around to the subject of his "very strange personality." Everyone agrees that it is not a type common in Japan: he is said to be openly ambitious. His supporters say that he is trying to bring about significant change in Japan and that he knows what he could accomplish if only he were given the means.

This man is vice-president of Tsukuba University and formerly the dean of science at Tokyo Education University. He was a student of Shinichiro Tomonaga, the 1965 Nobel Prize winner in physics, and he worked at the Institute for Physical and Chemical

Research (IPCR) in the 1930s under Yoshio Nishina.[7] He may be following a model established by Nishina:

> It is hardly thinkable—in prewar Japan, where family-style academic cliques prevailed—that a young man like Nishina, who had no position in the Imperial Academy and held no court rank or honors, could have planned and carried out such a major project [establishing his own lab within IPCR in 1931] on his own initiative.[8]

Tsukuba's founder's official position in the government is member of the science advisory committee. His influence in governmental circles is very strong, the physicists say, because his former college dormitory mates are now the leaders of the Liberal-Democratic Party, the conservative coalition that has governed Japan for over thirty-five years.[9]

Even if this man had had the personal power to establish the laboratories and university at Tsukuba in the form he envisioned, the physicists stress that the bureaucrats in Tokyo have the legal control of all funds, including the line-by-line items in the KEK budget. One theorist noted that when he wishes to publish his work, the bureaucrat in charge can tell him that the laboratory has no responsibility for publishing his "private" papers or for paying the postage for its delivery abroad. The physicist can get someone at a higher level to intercede, but he laments that when that bureaucrat gets transferred elsewhere he will face the same problem all over again. (Within the Monbusho agency that funds KEK, any bureaucrat can expect to be transferred to any number of positions during his or her career; even the administrators at various levels of KEK are employees of the government agency.) The physicist says that this difficulty in publishing his work has caused him to begin to question the usefulness of his work to society.

Another KEK physicist points out that in his office, which is used by seven research associates, there is only one telephone, and since no one likes to answer anyone else's calls most people tend to avoid the office altogether. In a community so dependent upon rapid communication, this typifies, he believes, the government's lack of sympathy for physics. He adds that those physicists who are influential with the government are theorists, whereas in the United States the physicists who influence science policy are the experimentalists. Furthermore, these Japanese theorists, in his opinion, tend to act and speak only on their own behalf, not as advocates for high energy physics or basic research.

KEK: A New Model for High Energy Physics Research in Japan

As a new institute differentiating itself in kind from INS, KEK must define its role with respect to the university physics departments in Japan. The departments wanted KEK to be organized as a resource for them, rather than as a powerful, independent force in the Japanese particle physics community. In the 1960s, when the design and structure of KEK were being debated, several physicists in the United States were formulating their proposal for Fermilab, the Fermi National Accelerator Laboratory, which was eventually built west of Chicago. Many of the Japanese and American physicists engaged in these plans knew one another well, and for some time the two projected labs bore a striking resemblance.[10] Many of the American physicists planning Fermilab were from university departments, and, as I have been told, they were often frustrated and annoyed by the advantages held by the resident physicists at the other national laboratories, especially SLAC and Brookhaven National Laboratory.

Fermilab was deliberately structured to give the advantage to the university physicists. Until recently physicists generally could not remain on the staff permanently at Fermilab as experimentalists unless they agreed to devote half their time to administration. Forcing administrative tasks on "in-house" groups at Fermilab served to prevent them from ever holding the power of the resident groups at SLAC or Brookhaven; most ambitious experimentalists would not accept these terms.

At Fermilab most research groups have been, in fact, short-lived amalgamations of "visitors" to the laboratory. (Each group usually has members from several different institutions, although some significant groups are a team from one institution.) When these "users" arrive at the laboratory to do their experiments, they occasionally find that other groups have been scheduled to do very similar experiments at the same time, which generates competition at the lab for the various facilities, such as the computer center. The leaders of users' groups often have other duties (teaching, in particular), and consequently they must commute to Fermilab to supervise their experiments. This can mean that it is the young people in the groups (graduate students and postdoctoral research asssociates) who are actually running the experiments on a day-to-day basis. All of these arrangements worked to enhance the power of the director and to establish a coalition between him and the users against the resident physicists. This organizational structure came under considerable scrutiny when the founding director of Fermilab submitted his resignation in protest against funding cuts from Washington and then privately asked to be considered to succeed himself. Another experimentalist was eventually named director, with the condition that the original administrative structure could be modified.

Given the power of established university departments in Japan one might expect them to find the original Fermilab model congenial. Furthermore, there was a precedent in the organization of the Research Institute for Fundamental Physics (RIFP), established at Kyoto for theorists by Hideki Yukawa after he won the Nobel Prize in 1953. Faculty members there are usually appointed for limited terms (three to seven years for full and associate professors, and one-and-a-half to four-and-a-half years for research associates).[11] I was told that when this system was introduced at RIFP it was successful—in part because jobs were more plentiful then, and in part because people were willing to take the risk in order to work with Yukawa. This organizational plan was then adopted at other scientific research institutes.

At the time of this study only users from universities had the official right to do experiments at KEK. All the resident physics groups were officially titled Facilities Groups, and as at Fermilab and SLAC the physicists in Facilities Groups were expected only to design and maintain the accelerator and detectors. They had limited access to research time. A senior physicist said that strong in-house experimental groups would be necessary at KEK, but he added that any such change would have to be achieved by the force of "strong personality." The physicists' appointments were for seven years. A young physicist at KEK said he expected this time limitation to be circumvented somehow. In fact there is now a mixture of permanent senior positions and temporary junior positions, not unlike the American postdoctoral research associates. One difficulty in establishing permanent research positions at KEK is the reluctance of senior physicists to leave their university appointments and the power that goes with them, because of the chair (*koza*) system. Funding from government agencies goes directly to the *koza*. Out of this allotment come salaries, research expenses, clerical staff, and janitorial services. In the sciences the initial funding for a *koza* will include money for a laboratory and equipment. All subsequent funding for that chair will only be adequate to maintain that equipment and the staff of the *koza*. This explains why physics departments at Japanese universities have specialized in one kind of particle physics: Nagoya on polarized targets, Tohoku on bubble chambers, Tokyo on counters, and so on.

It also explains why the university physicists want their research at KEK to be funded to their *koza*; they could then disburse the funds to KEK as necessary for the use and development of experimental equipment. This would entail *koza* ownership of any apparatus developed for their experiments (a practice which already prevails at INS). This arrangement also would give the university *kozas* considerable political leverage in the governance of KEK. Naturally, on the other hand, KEK physicists prefer that the funds be allotted directly to KEK. The Ministry of Education is not authorized to provide funds to private universities, except when a private university conducts research in an area not investigated by any national public university. The state and city universities are in an intermediate position. At the time of this study, the issue focused on applications by Tokyo Metropolitan University and Waseda University (which is private) for funding of research to be done at KEK.

Senior physicists in universities are also reluctant to leave their established *koza* for KEK because of the long-term ties within the university *koza*. A senior physicist at KEK said that he thought older faculty from Japanese universities would have great difficulty adjusting to KEK. At KEK there is rather more informality between people of different status; compared to Osaka, Kyoto, Sendai, or Tokyo, KEK physicists displayed what, to my eyes, was almost no pattern of deference.

The absence of an established structure of long-lasting working relationships at KEK (although the form of the *koza* is used) seems to have led to the emergence of leadership by "force of personality." A physicist at KEK said that a group leader there has difficulty leading in the usual familial manner because the groups have no tradition to provide cohesion. They are divided along political lines and also between those who have spent a significant period of time abroad and those who have not.[12] Furthermore, people who have worked in Europe tend to cluster together, as do those who worked in America, and so on. Many of the physicists at KEK, especially the younger ones, are not pleased to have leadership by "personality," and some are not on good terms with their leader. In a Japanese context this situation is very difficult, and it would not be attractive to a senior physicist established in a university *koza.*

In-house Groups and Users at KEK

In most academic fields in Japan, there is a system of just two university networks: this dual system has deep roots in Japanese history.[13] In high energy particle physics there are five, each identified with a distinct set of interpretations and style of working relationships. One is centered at Tokyo University, another at Kyoto University, and each of these has several satellites.[14] Tokyo University represents the establishment: that is, it has close ties to the Liberal-Democratic Party and Monbusho, and it is considered by outsiders to be conservative and elitist. Kyoto, Nagoya, and Hiroshima form an antiestablishment group, with a strong commitment to consensus decisionmaking; some of its members are associated with the Japanese Communist Party.[15] This group, although a minority in the Japan Physical Society (JPS), is said by outsiders to be able to control the policies of the JPS because it always votes as a bloc. Power within this network is achieved via "democracy" and "the multitude."

There is another grouping of "antiestablishment" people who are not necessarily pro–Kyoto-Nagoya-Hiroshima. Then there is a cluster of independent individuals who represent only themselves and stand outside the establishment debate. This group is composed of eminent senior physicists such as Hideki Yukawa and Tomonaga (whose daughter is married to an experimentalist at KEK). The fifth and final group is considered "feudal" by outsiders. This group stands outside the establishment-antiestablishment debate because of their commitment to hierarchically organized research groups with a strong man leading by force of personality. This position is identified with Tohoku University in Sendai. Two senior physicists at Tohoku have strong personal ties to Monbusho and, hence, could be called another kind of establishment group.

This five-part structure is replicated within KEK to some extent. The beam channel group has strong ties to Kyoto, the bubble chamber group to Tohoku, and the theory and counter groups to Tokyo. Since the beam channel group has less opportunity to do research, the influence of Kyoto is lessened at KEK. Counter physics currently has more status than bubble chamber physics, so the hegemony of Tokyo at KEK appears firm. The struggle between these groups makes policy decisions for KEK as a whole quite difficult. One physicist said that everyone wants to have both "democracy"—identified with the antiestablishment group—and "an amenable, agreeable manner"—associated with traditional, establishment organization.

The advisory committee for KEK is composed of twenty people, ten from within KEK and ten from other labs and university departments. The outsiders include five high energy physicists, two theorists, one cosmic ray physicist, and two nuclear physicists. These ratios suggest the influence of the JPS in overseeing work in particle physics.[16] The insiders group is composed of three *ex officio* members from the director's office, two from the accelerator department, two from the physics department (which includes the experimentalists), two from the technical department (including the computer group), and one from the theory group. The ten insiders are nominated and elected by their own group members. This process would be unthinkable in the United States, where more hierarchical, Tohoku-style methods prevail. The director usually confirms the outcome of the voting. While I was at KEK, the director denied confirmation to two people. In one case, I was told, youth was the problem; everyone at KEK was reluctant to elaborate. I learned eventually that the other person was a political radical; several senior people had moved to cancel his election because in their minds the man's very presence on the committee would be an implied insult to the government, and this, they argued, would endanger the lab itself. Others saw the move as an effort by the establishment groups to nullify the influence of the antiestablishment groups.

It was predictable that the problem of how to organize KEK would be solved by creating spheres of influence within the lab for the major constituencies in the particle physics community; but this system is fragile, because it forces groups that have no history of interaction into daily contact and collective decisionmaking.

Chie Nakane describes the relations of roughly equal groups in Japan as strained and difficult and argues that such groups, if unable to avoid interaction, will struggle to establish a hierarchical relation between them.[17] Relations between groups at SLAC can also be difficult; but they seem to manage to keep greater distance between themselves than the groups at KEK. This is no doubt possible because at SLAC the groups have no direct role in the governing of SLAC; that is accomplished by the director, who governs, in the Japanese sense, by "force of personality."

Working Abroad

The system is all the more fragile at KEK because of the presence of a group of people who are no longer fully loyal to any of the five positions outlined above: those who have studied and worked abroad. These physicists fall into three categories: those whose foreign experience came early in their careers, those who regularly collaborate with foreign physicists at laboratories outside Japan, and those who have permanent positions abroad. The first group is large, the second rather small, and the final group includes no more than ten to twenty people. Of course, those in the second category have usually also worked abroad early in their careers, and those who work permanently abroad are likely both to have worked abroad early in their careers and to have a history of collaboration with foreigners. Nonetheless, each of these three categories is regarded differently.

It is quite acceptable in many fields of work to be sent abroad by one's teacher or supervisor for further study. One's position in the *koza* or work group is not altered by these arranged trips. Furthermore, it is an established practice among experimental particle physicists for a group taking on a new project to send younger group members to other groups, where they will acquire specific skills and experience needed by the home group.[18] These visits are arranged informally by senior Japanese physicists and group leaders abroad. For example, in 1976 one young man from KEK was working on an experiment at CERN, in Switzerland, because a senior scientist from the United States visiting KEK had told a group leader at KEK that he could use a young physicist on his next experiment at CERN. The Japanese who work abroad, and senior Japanese physicists visiting foreign labs, also often serve as contacts for placing young physicists for these sojourns. In some of the groups at KEK there is a clear sense that all the young people should have the opportunity to work on an experiment abroad. Some of the groups in the Kyoto-Nagoya-Hiroshima network, though, definitely do not encourage work in Europe or the United States. Everyone I asked in Japan knew which members of their group (and often many others) had been abroad, where, when, and for how long.

One young physicist at KEK spoke of his two years' work in a European lab in a way that I heard echoed many times over. He was impressed by the lab's affluence and its highly developed network of communication, including the large number of short-term visitors. But he was also disturbed by various events he observed. He noted that many students who were very bright and talented were forced to take positions in industry because their professors had taken a personal (not intellectual or political) dislike to them, a dislike that appeared quite arbitrary and unscientific to this young Japanese. He was incredulous at the injustice he perceived. He was startled by the power of the group leaders. New projects were adopted only if a senior physicist chose to become involved. Decisions were made by senior people alone, and younger people were informed of these decisions only if they were on close terms with the decisionmakers. Since his return to Japan, he has appreciated more fully the freedom, responsibility, and independence granted young physicists there. He also finds it much easier to do physics in his native language. But he does miss the affluence, communication links, and the experimental physics experience that the foreign lab offered. He is eager to see these advantages developed in Japan, and he has a strong expectation that they will be.

Older Japanese physicists who have worked abroad for longer periods and who have collaborated on work done at various foreign labs feel that the Japanese particle physics community is not a good environment for someone who is ambitious and independent. They say it is difficult to pursue one's own ideas, even if one is a group leader. One senior physicist said bluntly that he did better physics abroad. The physicists who regularly work abroad also face criticism from their colleagues. I was told by several scientists that working abroad was obviously very profitable in the short run for the few individuals involved, but that it contributed very little to the development of the Japanese particle physics community as a whole, especially when funding was limited. Many of those who have worked abroad are identified as having "strong personalities," and there is doubt in most cases whether these will be put to use serving particle physics in Japan. Because of these attitudes, many of those who regularly work abroad cluster together in the same research groups when they are back in Japan.

The third category includes those who have permanent positions abroad. I learned from a Japanese who has worked in the United States over twenty years that under new regulations the Japan Society for the Promotion of Science (Nippon Gakujutsu Shin-Kokai, similar in mission to the National Science Foundation in the United States) can use their foreign visitors' funds to invite Japanese nationals back to visit if they have been abroad more than ten years. Several physicists in Japan pointed out that this arrangement was very rewarding for those who had remained in Japan because there was no language barrier. These visitors were seen as making valuable contributions by serving as an active communications link. Some Japanese working abroad have been invited to return permanently to KEK, where they would be given high positions and much freedom. One Japanese physicist did return, giving up a tenured position in the United States, but left again after one year.

The language barrier is an important topic for the Japanese. American English is the *lingua franca* of their field. At many universities in Japan graduate students must present their work in English in certain seminars, and some articles written by Japanese are published in Japan in English.The Japanese are aware that they must be able to communicate effectively with foreigners in American English for their work to have the recognition they very much want. Many physicists sought me out to discuss issues so that they could practice speaking English and learn the American English terms in current usage among foreign physicists. Two asked me to check the translations of articles they had written. As a result, my own Japanese language skills did not develop quickly—but on the other hand, I gained access to many people.

The many physicists at KEK who have worked abroad have a partly independent perspective on the five-part organization of the Japanese particle physics community and its replication at KEK. They do not have enough influence or status to negotiate a compromise between the five factions.[19] Nonetheless, their frustration with the problem suggests that there are other approaches.

The physicists want to know not only "the best environment for good physics," but also what environment is best for the development of a strong experimental particle physics community in Japan. All are agreed that Japan is currently an underdog in the international standings. Some groups argue that the insiders at the lab must be strong, that this is the traditional organization of academic research in Japan. Others, mostly those trained abroad years ago, insist that strong, well-funded, university-based groups should expand their collaborations at foreign universities. Younger physicists, many recently returned from abroad, sorely lament the state of resources in Japan, and insist that the proper path is to fund a variety of labs in Japan so that the next generation of physicists in Japan can acquire the broad experience necessary to do first-rate physics without having to work long periods outside their own country. They argue that as long as young experimentalists can broaden their training only by going abroad, where much of their energy is spent adapting to a foreign culture, there will not be a first-rate particle physics community in Japan.

In the midst of this debate people at KEK are in the process of creating the facilities, adjusting the formal organization, deciding who will have access to the lab, and negotiating with the government about long-range funding.

Contrasts between Japan and the United States

At both KEK and SLAC, new facilities were funded by the respective national governments with the specific injunction that they be open to all qualified users. Both the Japanese and American particle physics communities need to find ways of balancing the interest of insiders and users (usually university-based) in the management and use of important new facilities. The physicists are being forced to reevaluate what is the best organizational environment for physics. This problem, which confronts the entire particle physics community internationally, is part of a larger issue: the institutionalization of a new and very expensive field of inquiry within a highly conservative university tradition and its impact on the established allocation of research funds.

Their different approaches point to important distinctions between the Japanese and American communities. In the physicists' own opinion their differences are due to their disparate ranking in the international particle physics community: SLAC has been "at the center of the action" and is trying to maintain that position; KEK, although ambitious, is a comparatively modest institution. In my view the differences between Japanese and American practices in laboratory organization and in training new physicists are due not mainly to the relative sophistication of the technology, but to the culture of the physicists. American particle physicists differ in many ways from their Japanese counterparts, and these differences correspond to strong cultural values. They can be grouped under the headings of teaching, group and laboratory organization, leadership styles, inheritance, and historical perspective.

As I present the "Japanese" and "American" styles I remind the reader first that I am describing what Max Weber called "ideal types." These are abstractions, for the sake of analysis: no one person or group would fit the model perfectly. Second, it is crucial to keep in mind that *both* models exist in *each* country. What I will describe as the dominant "Japanese" style also exists in the United States; and the prevailing style in the United States most certainly also exists in Japan.

The Japanese perception of how Japan ought to go about becoming a first-rate particle physics nation places great emphasis on training the next generation. This transfers to the young physicists great responsibility for the future, and also confers upon them great obligation to their teachers. This mutual responsibility and obligation is consistent with the tenets of *amae* (interdependence, particularly in generational ties), a crucial value in Japanese culture.[20]

Americans believe that what is best for the future of physics is for each physicist to provide the best physics possible; what is good for the individual will benefit the community. I discussed earlier how the educational process in physics cultivates this individualistic, competitive drive in each student. This value is consistent with the American cult of *individualism*, a laissez-faire economic model for the growth of knowledge: each contributes as one chooses, and the unfettered marketplace of ideas selects the best. In this model, self-interests are seen as properly and necessarily competing. Not only Japanese but also Europeans find these assumptions alien: one European working at SLAC said that Americans have no understanding of collaborative intellectual exploration, because they see ideas as real estate—that is, private property.[21]

In Japan when I asked group leaders where new ideas for experimental design or data analysis come from, they generally credited the graduate students and *koshi*; they said the group then responds to their ideas, perhaps modifying or amplifying them.[22] In the United States, the senior physicists say that it is the leader who generates ideas, which the group amplifies and executes under his direction. While group leaders each will privately acknowledge that "postdocs often pull me out of the fire," they stress that this must not be conveyed to the postdocs or anyone else in the community. Each speaks as if he were the only one who has experienced this.

Japanese groups see themselves as much more democratic and less hierarchical than the American groups. The Japanese who have worked in the United States indicate that they were quite surprised by the formality and structure they found there, and they wonder if this is really the best environment for physics. In Japan, the groups make their decisions by consensus, even if they sometimes find the process tedious. It is said that the consensus model is closely adhered to in the southwest, especially at Nagoya, and less so as one moves northeast toward Sendai, where a more "traditional" style is said to prevail. In all the Japanese labs I visited, I found everyone actively discussing various matters of laboratory-wide concern before the decisions were made. All seem to feel themselves well informed. Japanese values direct the leaders, even the most feudal, to consult fully with all those in their groups and, in turn, require all members to advise thoughtfully. The leader is to decide on a policy consistent with both good judgment and the health of the group and its members. In accordance of this ethos of *wa*, it is the responsibility of group members to cooperate with the decision *if* they believe this process has been respected.[23] The demonstrations at the site of the Narita airport were a case in which a community rebelled—and are now renewing their campaign—because they felt that they had not been properly consulted.

In the United States, although the style is informal, the group structure is hierarchical. Every decision is made by the group leader, who then informs the group of how it is to be implemented. At SLAC when I asked postdocs what they thought of certain issues confronting the lab, almost all said that they did not know what was going on. A few said bitterly that "they only tell us what's happening after they decide." Often they would question me, saying, "You seem to be better informed than we are."

Those at each level of the hierarchy are expected to observe and listen to those above and pattern their behavior accordingly; it is not appropriate to comment negatively on those in positions of greater status, no matter how informal the relationship. Informality is a gift or reward bestowed by those in charge. Competition is a fundamental tenet of American culture; it is very difficult for most Americans to imagine that competition has any negative implications or that there could be any other motivation for performing with excellence. Where competition is unfettered, as it is supposed to be in the American particle physics community, organizational hierarchy is seen as a *natural* ranking of human talents. The leader is the best and most entitled to judge—otherwise, he would not have become the leader. There is occasional discontent among those excluded from the decision process, but there is no self-doubt among the leaders.

In Japan there is no strict division of labor among the physicists in the lab. They feel that this makes for better experimentalists, with a good understanding of the entire experimental process involving their type of detector. Japanese physicists have in general a wider range of responsibilities than their counterparts in the United States: for example, since so much of the research apparatus is constructed by private industry and then assembled at the lab, experimentalists must be able to design what they want, to choose a manufacturing firm, and to communicate with its representatives in fully explicit detail; there is much less room for trial and error in detector design. And because of the difficulties in attracting trained technicians to a lab, experimentalists must train the technicians themselves, even in the most routine tasks, and closely supervise their work.

In Japan one "cultural" model for a work group is the *ie*.[24] Generally speaking, this refers to an entire household, including its privileges and responsibilities in the network of households of which it is a part. It is the obligation of each member of the *ie* to keep the household and its resources intact and pass them on to the next generation. Status in the *ie* is determined by age, not by competition, there is no strict division of labor, and all members contribute their point of view in discussions about decisionmaking and planning. The organization of most particle physics research groups in Japan is consistent with this model of the *ie*.

In the United States a major "cultural" model of a work group is not a large household, but a sports team.[25] That is, the leader is like a coach directing a team of football players, each of whom has specialized, distinctive skills. The coach is the only team member to understand the entire game process and the only one empowered to design the team's strategies and tactics. The membership and organization of the team survives only as long as it is "winning," accumulating a better record in competition with other teams. If the team consistently loses, the owners will disband the team and assemble a new one, perhaps even relocate to a new city. Any particular game may include, for instance, players and coaches who have worn the other team's uniform; people on different teams may have been teammates once. Obviously, this process generates very strong professional loyalties and very weak institutional ties. This team model is congruent with the organization of research groups in American particle physics, as well as of some of the Japanese groups.

At SLAC the problem of technicians does not exist; they are widely recognized in the community as doing excellent work, often on their own initiative, often quite innovative. Some groups have tried to lure technicians away from one group to another but it is a subtle and complicated process. Almost all of the research apparatus, including the accelerator, is constructed at SLAC. Physicists in the groups work closely with the technicians to build and assemble the parts of their group's detector. There is a rather strict division of labor among the physicists; it is said that only newcomers and the group leader have an overview of what is developing. Curiously, at other American laboratories (such as Brookhaven) where there are fewer highly skilled technicians and sections of the research apparatus may be constructed outside the lab, this same strict division of labor seems to prevail within the group.

Another difference concerns the amount of experience the experimentalists have in working with a variety of detectors. In Japan, research groups are very much identified with a specific kind of detector; this is generally a consequence of the patterns of funding from Monbusho and of the *koza* system. Because of the endogamous recruiting system, once students enter graduate school they have effectively delimited the kinds of research they can do for the rest of their careers. Of course, the establishment of KEK is complicating this somewhat, as does the opportunity to study and work abroad for long periods. In Japan, great emphasis is placed on purity (*wabi*), as opposed to contamination. Blends, amalgamations, and mixtures are not valued at all. To be an unmixed product of one school of thought (*iemoto*) is to display self-discipline and commitment; this is the proper context for achievement and creativity.[26]

By contrast, in the United States, postdocs are typically encouraged to vary their experience by working in areas of physics different from their concentration as students. Some groups have worked, as a whole, on different detectors. Many senior experimentalists seem to have deliberately set about working on a variety of machines before they propose the construction of a new detector of their own design. Americans value independence highly, and consider mobility—exposing oneself to many influences—a sign of independence. Intergenerational allegiances come into force only at the end of the postdoc years and then exogamously—a more lasting loyalty is felt to the mentor than to the group leader.

Although the Japanese groups and networks are fixed and recruits to new positions must come from within and the Americans insist upon movement through groups, especially at the early stages of one's career, the Japanese community seems to tolerate much more diversity in leadership styles and group organization than the Americans. This may be the result of the internal recruitment: one need not adapt to another group's style. The Americans move frequently and expect more similarity. Nevertheless, I would argue that there are two main styles of leadership in each country.

In American universities in general taking a position in the department in which one was trained is considered extremely bad judgment by both the student and the department. It is presumed by others that the student could not be placed elsewhere, either because the student's work was inadequate or because the faculty network ties were weak. The only respectable way to take a position in the department of one's training is to go to another school, establish an independent reputation, and then return. As everyone says, "you must go away to come back." Independence is crucial, in this value system, to achievement and creativity of any kind; indeed it is felt to be a condition of self-discipline and intellectual commitment.

Another contrast, which was discussed in Chapter 3, concerns the opportunity for senior physicists to bequeath to younger physicists their positions, networks, equipment, such as detectors, and access to scarce resources, such as funding, beamtime, and computer time. For the Japanese these are bestowed by the *koza* leader to his successor; in the United States they must be both redistributed and reconstituted in each generation, perhaps even more often.

The most striking distinction of all between the Japanese and American particle physics communities is in their awareness of historical change. In Japan they assume that they are "underdogs," on the periphery of power, internationally, and they are trying to change that.[27] The Americans with whom I talked approach the core-periphery question with complacency. They agree with the Japanese that the United States is "where it's at."

Americans see this as a contest and they are committed to staying on top. Only in winning is one's effort vindicated; to be second is to be a loser. In Japan to be second is not to be a loser, but to be a younger brother, an honorable role and often a preferable one. The Japanese do not seem to regard their movement toward the core as a contest, but as part of a rather predictable historical process of new scientific communities emerging alongside older ones. At the very least they think it is possible to narrow the gap between their community and the Americans. The Americans apparently believe that their own status in the current international ranking is unchallengeable, unless they are crippled by external events like funding cuts. This lack of historical awareness of change among the Americans, and the acute consciousness of it among the Japanese, is also demonstrated in their different explanations of the decline in funding and quality of students in particle physics.

Current conflicts within the Japanese and American particle physics communities expose the latent instabilities in both. In the American case one such conflict is caused by new pressures to incorporate "client groups" (in anthropological terms) as opposed to individuals into a "residential cluster," incorporating user groups rather than individual novice physicists into the day-to-day organization and management of a laboratory. Incorporating user groups into the management of SLAC disturbs the informal but hitherto highly effective balance within the lab, creating a need for extensive political authority to be concentrated formally in the hands of the director, who consequently no longer merely arbitrates or leads by force of personality. The director no longer is first among equals; his role is now that of a chief and all aspects of social relations are in the process of transformation.

For the Japanese the establishment of KEK as a technically advanced national laboratory means that university groups must circulate through KEK in order to maintain their own power. Groups traditionally kept separate could cultivate differences in styles of leadership and organization; but in moving together at KEK they find that their differences are incommensurable and they cannot work together. In response they are relinquishing their diversity and their exclusive loyalty to their chief and beginning to turn to those who are most adept at dealing with another system of organization: those who have worked abroad. The funding environment for each system is changing, leading to instabilities. That is, scientific funding is now expanding in Japan and leveling off in the United States. The research organizations' structures and leadership styles are no longer appropriate in their new funding environments.

One solution would be for each community to adopt the other's organizational structure and leadership styles. As I have said, examples of both exist in each country: the "coach/team" approach can be found at national laboratories in Japan and the "*sensei/iemoto*" (esteemed teachers/home school) approach can be found in the United States at a few universities, such as MIT, and perhaps even at a national laboratory like the Stanford Linear Accelerator Center. It remains to be seen if each country's scientists will shift away from their dominant styles to the less common model. Will "coaches" who have learned to locate players with highly developed skills and to design strategies for winning every encounter with other teams now begin to build long-lived groups and designate successors? Will *sensei* accustomed to maintaining *ie* across generations and defining the boundaries between *iemoto* now begin to form ties with other *sensei* outside their traditional networks and encourage the new recruits to spend time working in various *iemoto*? Will they allow the next generation to name their own leaders and form their own groups?

Collaboration

The cultural differences between the American and Japanese particle physics community and the recent shifts in the nature of their differences can be observed at work in international interactions. Toward the end of my fieldwork, a Japanese group and one of the groups at SLAC began to arrange a long-term collaboration. The motivations of the two groups were clearly distinct. In recent years, the American and European physicists had suffered from funding declines. Negotiations between the Japanese and United States governments during the administrations of President Carter and Prime Minister Fukuda on the subject of trade imbalances led to significant funds being made available by the Japanese for Japanese-American collaborative research on energy, to be conducted in the United States. One of the group leaders at SLAC realized that he could supplement his group's budget with these new funds: the group's membership would be enlarged by Japanese physicists whose salaries would be paid by the Japanese government, and new components for the group's detector could be constructed, although they would remain the property of the Japanese group. The leader was gaining funding on his own while avoiding negotiations with other leaders and networks in the United States, within and outside the lab, and the concessions such negotiations would entail.

The Japanese group was from a university department with firm ties to INS and its leader was "strong and active." Following traditional patterns, the leader could establish a new international network tie with the collaboration, which would later allow him to send many of his students and *koshi* abroad for advanced training. He was also following a model established by Fermi in the 1920s and 1930s for building a strong group: gain expertise abroad in specific skills, and then use these at home to construct one's own detector. Furthermore, the *koshi* would build personal ties for international leadership in the next generation for their *koza*.

Both group leaders tried to specify their plans and expectations in writing. Such written contracts are unheard of in the United States and, as far as I know, also in Japan; but in this case, the Japanese physicists in particular came to think that a written accord was appropriate. The Japanese group leader spent two weeks talking to many people at SLAC and then said bluntly to the SLAC group leader that he would have greater confidence in their agreement if it were written. Collaborations at SLAC, until the conflict over PEP, had been structured on the model of the strong in-house group and weak user group. The Japanese group leader was restructuring that model still further by insisting on a written agreement. The Japanese group's subtle and thorough investigation of the American group's strengths and weaknesses as seen by American physicists displayed both an astute sense of how to gather such information and a confidence in their own bargaining power. The Japanese leader's extensive consultations and his desire for an open agreement, fully explicit and widely known, were consistent with Japanese practices; he was also seen to be in frequent communication with his home base.

The SLAC group leader signed the agreement he and the Japanese leader wrote. He then showed the document to SLAC's director. They were both nervous about the implications of the written agreement. They decided together that the document only meant that the group leader and the lab "had agreed to agree" with one Japanese, and that the legal dangers and uncertainties were compensated for by the large funding that would come to the group. In sharp contrast to the Japanese practice, these two men made the decision for SLAC. The proposals for funding made by each group to its respective governments were rather different, in spite of the written accord. Each group examined the other's funding report carefully. As the SLAC group leader said when showing me the Japanese proposal, "I know how to read between the lines with Americans and Europeans, but this is different. I don't understand." This same group leader had told me early in my fieldwork that culture had no place in the international particle physics community.

As they tried to work out their collaboration, it had become apparent to both group leaders that they could not rely on shared assumptions and expectations. But they did not identify their problem as a cross-cultural misunderstanding. Both saw themselves primarily as members of the particle physics community and consequently expected to have the same views about how to arrange a collaboration. When they found this was not the case, each began to feel distrust for the other. The Japanese group leader dealt with his suspicion by proposing a written contract. The SLAC group leader agreed, but his distrust was expressed in the close readings that he, the lab director, and SLAC's business manager gave the contract. The gap between them widened when their separate funding proposals emerged, and I was asked "to read between the lines" of the Japanese document. The gap between the lines was the gap between Japanese and American culture. What particularly troubled the SLAC physicist was that the participating Japanese physicists were not listed in order of their reputations. He feared that the Japanese group leader planned to use the physicists in the order they were listed, which could only mean that he was not seriously committed to the collaboration. I showed him that the physicists were listed in order of the rank of their institutions, and within that, according to their age. I told him that this order would impress Monbusho and show professional "savvy," just as ordering by reputation would impress SLAC's program advisory committee. Institutional affiliation was implicit in the explicit American ranking of reputations; the ranking by institutions in the Japanese proposal displayed an understanding of traditional forms. It was assumed in the Japanese document that readers would know informally the reputations of the specific individuals. Each document was straightforward and predictable within the set of shared understandings understood by the author; each was trustworthy within its own culture.

Fortunately, the two physicists also share a culture—the culture of the particle physics community. They both realized that the Japanese group leader was dealing with the collaboration as a negotiation between two strong groups to enlarge a detector. They also knew that the SLAC group leader saw it as an arrangement between an in-house group and a user group for the users' temporary access to a detector he controlled. They also knew that the SLAC group leader saw their "top dog" status verified; the Japanese group accepted their "underdog" status, for the time being. They both understood that the detector itself was central.

The immediate issue confronting the Americans and Japanese in changing subsistence ecologies is how to incorporate new facilities for doing physics, such as PEP and TRISTAN, while maintaining the stable structure of their communities. The Japanese and American collaborators were each preparing their own group to be in a better position to compete for effective use of an anticipated facility at their respective laboratories. The SLAC group wanted to build a new detector at PEP and the Japanese group planned to build one at TRISTAN. Each group was garnering from their collaboration different resources to use in their competition with other groups in their own country. Nevertheless, disputes in the collaboration continued to be formulated in terms of the "in-house" and "user" relationship, as understood by each group. This extended even to the decision regarding final possession of the complete set of data (on computer tape) collected in their experiment. The SLAC group wanted to maintain the prerogatives of an "in-house" group at SLAC: the Japanese insisted on their expected rights as users to data they had not only helped generate, but also helped pay for the equipment to collect.

The high energy physics communities in each country are debating the model of a good laboratory as part of a larger question: how ought the physics community be organized to pursue physics in the future, as well as in the present? Neither of the experimentalist groups in this story expected at the beginning of their collaboration that their long-term goals would require them to continue negotiating after their experiment concluded. When the Americans arrive at KEK they will discover the subtle and significant differences in the Japanese formulation of the debate about the best environment for physics and how this affects the relationship between "in-house" and "user" groups. That is, they will be confronted with a different formulation of how a national high energy physics community ought to be organized within the context of the larger international community. Leadership in international physics remains the prize; each country's groups would prefer to see their own teams win. Each side acknowledges the general structure of the enterprise, but strategies and tactics will differ significantly and according to each group's cultural repertoire.[28]

EPILOGUE

Knowledge and Passion

In this book I have examined the high energy physics community: the organization of the community, the stages of a career within it, the physical theories its members share, and the environment and machinery physicists build in order to do their work. Anthropologically speaking, I have described their social organization, developmental cycle, cosmology, and material culture. I have explored a theory originally formulated by Durkheim and developed in many ethnographies over several decades, a theory which proposes that a culture's cosmology—its ideas about space and time and its explanation for the world—is reflected in the domain of social action.[1] In other words, ideas about time and space structure social relations, and the spatial and temporal patterns of human activity correspond to people's concepts of time and space. I have found this notion fruitful in studying the high energy physics community: their physical theories of time and space *do* illuminate their social reality.

The social practices of particle physicists as well as their explanations of the physical world display vivid contrasts among different kinds of time. Physicists use this knowledge, which is the product of their activity, to form a single taxonomic system, with unchanging laws of nature, free of ephemeral time. Yet they see the process that generates this very special product as one of linear, progressive accumulation of knowledge, to which individuals can make more or less illustrious contributions. Their everyday working lives are organized in terms of several different, intense timescales. There is the precious, negotiable commodity of beamtime, which can be accumulated and bartered, and, on the other hand, the very limited and irretrievable lifetimes they can foresee for their labs, their machines, their ideas—and for themselves.

The tension in laboratory life between the desired static product and the ineluctable, unforgiving process of making it is similar to the difference between relativistic time, in which the particle interactions take place, and nonrelativistic, experiential time, into which these interactions must be transcribed by detectors before their message can be delivered to human senses. These two kinds of cosmological time converge in detectors. Detectors express diverse styles of research, different strategies for success. Above all, they embody rules for separating noise, which arises from imperfections in the experimental process, from data, which carry authentic messages from nature. Good detectors are also, by definition, imperfect devices for receiving signals from nature; nevertheless, detectors ideally become invisible, transparent scientific instruments for reading news from nature. I see detectors as the material embodiment of the high energy physics culture. They display the tension inherent in the physicists' different experiences and ideas about time and they also provide a model for the resolution of that tension.

Detectors and Desire

The relationship between scientist and nature is at its most intimate and physical in the detectors, the sensitive research devices that are constructed to receive the signature of nature. These apparently passive machines are like a trap set to snare elusive signals from a capricious nature. The language used by physicists about and around detectors is genital: the imagery of the names SPEAR, SLAC, and PEP is clear, as is the reference to the "beam" as "up" or "down." One must see the magnets at LASS to appreciate the labial associations in the detector's name, Large Aperture Solenoid Spectrometer. Ironically, the denial of human agency in the construction of science coexists with the imaging of scientists as male and nature as female. Detectors are the site of their coupling: standing on the massive, throbbing body of the eighty-two-inch bubble chamber at SLAC while watching the accelerated particles from the beam collide twice a second with superheated hydrogen molecules made this quite clear to me. The energy of those collisions was transformed, sometimes generating particles never seen before. The consummation of the marriage between scientist and nature in the detector sometimes leads to progeny for the proud scientist: a discovery, attesting that he is a real scientist.

Those progeny will be named a discovery only if they can be reproduced, if their production can be endlessly repeated in other detectors, given the proper coding. It is the job of the scientist to identify that coding, to show that the traces of nature in the machine are not noise, but data. Providing the correct reading of those traces will enable the making of the same traces elsewhere. In this way, the physicists prove that their reading, their physical theories of the traces, works.[2] Nothing else establishes the realism of these interpretations. As theorists say, "If I weren't interested in having my ideas proven real, I'd be a mathematician." Theorists need the detectors to gain that realism. The difficulty is in finding that one reading which works. The method for finding that one real reading is in knowing how to make cuts such that one is left with good data on the one hand and noise on the other, much as one learns to make cuts among novices, leaving only scientists. Knowing where to make cuts, according to the experimentalists, means knowing one's detector, and the only way to know a detector is to build it.

If the cuts are made properly and the remaining data are significant and reproducible in other detectors, especially other kinds of detectors, then the data constitute a discovery. The meaning of the word *reproducible* here is problematic: no two detectors are alike. No one could get funding to build a copy of another detector and no one would want to try: there would be no credit and influence to be gained.[3] Furthermore, only the group that built the original would have the knowledge to build the copy. If one must know a detector deeply in order to strip its data of noise, one must also be able to convince one's colleagues that the detector did not make the data. One must argue convincingly that the detector is indeed only a transparent scientific instrument that passively and objectively recorded information about and from nature.

Yet for the experimentalists themselves, the detectors never cease to be very expensive congealed human labor. It is the product of the group, its representation, and its signature, just as the data in the detector is presumed to be nature's representation, and signature. Experimentalists read their detectors not only as records of nature, but also as mnemonic devices for the history, present, and future of the research group. The detector is the visible sign of their skill and talent as scientists.

Theorists themselves have little access to, or knowledge about, the detectors. When experimentalists present papers at seminars and conferences, they always begin with a detailed description of their detector and devote at least a third of their talks to these machines before introducing the data generated in their experiments and reporting how those data were analyzed in order to produce "curves" (interpretations which have an acceptable degree of "fit" with the data). Nevertheless, the theorists rarely attend carefully to these "technical details," referring to them as the "scotch tape" part of the talk. It is the theorist who is more likely to see detectors as scientific instruments which simply record nature, as transcription devices which themselves leave no trace. The rather Platonic theorists, unlike the Cartesian experimentalists, see the data produced by the detector as being uncontaminated by the machine.

Text and Authority

Heinz Pagels has said that "nature is the text, a cipher to be deciphered."[4] The intellectual roots of the image of nature as a text lie in biblical exegesis, with the assumption that the Bible is the written word of God and, by analogy, that "the book of nature" is the manifestation of God's purposes. To scientists, especially when they give historical accounts, their machines are like eyeglasses through which they read and then decipher the fixed text of nature. The physicists see data and nature as equivalent. Experimentalists want to see themselves as the decoders, or at most as the ghostwriters, of a story whose original author is nature. Theorists want to see the data produced by experimentalists with the help of the machines as a text directly authored by nature.

For Descartes the media for acquiring information about nature were the senses, which he recognized as flawed. But he believed that there was a guarantee that what human minds apprehend from this sense data actually corresponded to nature and was not mere illusion; his guarantee was God. Galileo and Bacon believed that scientific instruments could assist God in assuring the validity of one's data. Presumably, machines could reproduce data which would resemble that of the senses, but without the inherent subjective biases of the biological senses. Just as God was the mechanism by which Descartes assured himself of the veracity of his subjective sensations, machines now affirm the validity of the physical theories and interpretations of the particle physicists: the role of human agency in the construction of scientific facts is denied.

Latour and Woolgar, describing laboratory practices in immunology, also treat data as a text, but they credit the machine as its originator and author; for them scientists are astute decoders of machine-generated data-texts.[5] In my account machines themselves are the texts, variable texts. The physicists are engaged in the incessant production and reading of machines in which neither the machine-text nor their reading of it is ever fixed. These texts and readings are imbedded in community traditions about how to interpret nature and identify discovery. Reading the detectors has enabled me to tell a story about high energy physics culture; reading machine-texts enabled me to describe the reproduction of nature, the construction of discovery, and the reproduction of physicists in this community.

Common Sense

Some readers of my earlier work have thought that I have represented these scientists as producing merely fictions. I am not trying to evaluate the truth claims of the physicists' statements about nature. The first job of an ethnographer is to engage in a willing suspension of belief in the subjects' commonsense world. We may be able to accept that suspension easily when the anthropologist is describing some "primitive" people in an "exotic" setting or some "backward" culture, remote from our own world view. Our facile acceptance of the ethnographer's suspension of belief in the world view of her subjects in those cases can conceal simple, perhaps unconscious, condescension. When the ethnographer brings this same analytic distance to the study of our own society, it can be very disturbing. The "common sense" of our world is buttressed by a profound belief that it is "scientific," which means that it is "true." Or, as the physicists whose sense of certainty about this world view I have been studying would say, "it must be true because it works." Exploring the correlation between our efficacy in the world and our explanations of that efficacy is not my task.

There are close interconnections between the belief system of the community under study here and the common sense of the culture in which the ethnographer and most of her readers were raised. Maintaining suspension of belief under these conditions is a delicate business and subject to pitfalls. The ethnographer is heard (by readers and informants) as calling into question things that "stand to reason," and this stance can be read as a hostile or subversive skepticism. There is no guarantee against such readings: the ethnographer can only watch her language.

I have never met a high energy physicist who would entertain for a moment the question of whether electrons "exist" or not; and I can sympathize with that, for unlike some of my more reflexivist colleagues, I find it appropriate to assume that physicists exist. Unlike most physicists, though, I do recognize the importance of the question, in a less abrupt form: where do the social categories of physicist and physics community and physics culture exist? I mean this book to address that question. I have presented an account of how high energy physicists construct their world and represent it to themselves as free of their own agency, a description, as thick as I could make it, of an extreme culture of objectivity: a culture of no culture, which longs passionately for a world without loose ends, without temperament, gender, nationalism, or other sources of disorder—for a world outside human space and time.

NOTES

INDEX

Notes

Prologue: An Anthropologist Studies Physicists

1. "A Theory with Strings Attached," *Time*, November 24, 1986, p. 52; Freeman Dyson, *Disturbing the Universe* (New York: Harper and Row, 1979); Richard P. Feynman, *"Surely You're Joking, Mr. Feynman!" Adventures of a Curious Character,* as told to Ralph Leighton, ed. Edward Hutchings (New York: Norton, 1985); Hideki Yukawa, *Tabibito,* trans. L. Brown and R. Yoshida (Singapore: World Scientific Publishing Co., 1982); Roland Barthes, "The Brain of Einstein," *Mythologies,* selected and trans. Annette Lavers (New York: Hill and Wang, 1972), pp. 68–70; Daniel J. Kevles, *The Physicists: The History of a Scientific Community in Modern America* (New York: Alfred A. Knopf, 1978; Cambridge, Mass.: Harvard University Press, 1987).

2. Kevles, *The Physicists* (n. 1).

3. Paul K. Hoch, "The Crystallization of a Strategic Alliance: American Physics and the Military in the 1940s," paper presented to the Sociology of Science Yearbook Conference on Science and the Military at Harvard University, January 1987.

4. For surveys of the history of particle physics, see Albert Einstein and Leopold Infeld, *The Evolution of Physics: The Growth of Ideas from Early Concepts to Relativity and Quanta* (New York: Simon and Schuster, 1938); Richard Feynman, *The Character of Physical Law* (Cambridge: Cambridge University Press, 1965); Walter Fuchs, *Physics for the Modern Mind,* trans. M. Wilson and M. Wheaton (New York: Macmillan, 1967); Werner Heisenberg, *Physics and Philosophy: The Revolution in Modern Science* (New York: Harper, 1958); Mary B. Hesse, *Forces and Fields: The Concept of Action at a Distance in the History of Physics* (London: T. Nelson, 1961); Gerald Holton, *Thematic Origins of Scientific Thought: Kepler to Einstein* (Cambridge, Mass.: Harvard University Press, 1973); Stanley Jaki, *The Relevance of Physics* (Chicago: University of Chicago Press, 1966); Sir James Jeans, *The Growth of Physical Science* (Cambridge: Cambridge University Press, 1950);

Stephen Toulmin and June Goodfield, *The Architecture of Matter* (New York: Harper and Row, 1962).

5. On the notion of "invisible colleges," see Diana Crane, *Invisible Colleges* (Chicago: University of Chicago Press, 1972). The numbers mentioned are those used repeated by informants; I have done no survey myself. In other words, these numbers represent the informants' sense of the size of their community.

6. Kevles, *The Physicists* (n. 1). For a less flattering view of Lawrence (a perspective rarely taken in print), see Robert Hermann, Letters Column, *Physics Today,* November 1977.

7. On new directions in anthropological research, including "repatriated anthropology," see George M. Marcus and Michael M. J. Fischer, *Anthropology as Cultural Critique: An Experimental Moment in the Human Sciences* (Chicago: University of Chicago Press, 1985). Benedict, *The Chrysanthemum and the Sword: Patterns of Japanese Culture* (New York: New American Library, 1967), was originally published in 1946.

8. See Charles Brazerman, *On Rhetoric in Science* (Madison: University of Wisconsin Press, forthcoming).

9. For an introduction to current debates in anthropology on the form and content of ethnographies, see *Writing Culture: The Poetics and Politics of Ethnography,* ed. James Clifford and George Marcus (Chicago: University of Chicago Press, 1986).

10. See David Schneider, "Notes toward a Theory of Culture," and Clifford Geertz, "From the Native's Point of View: On the Nature of Anthropological Understanding," in *Meaning in Anthropology,* ed. Keith Basso and Henry A. Selby (Albuquerque, N.M.: University of New Mexico Press, 1976).

11. Important research on the social construction of (contested) scientific and technological knowledge is often published in the journal *Social Studies of Science* and I recommend it highly. This research is founded upon the insights in two crucial books whose importance cannot be overemphasized: Paul K. Feyerabend, *Against Method* (London: New Left Books, 1975), and Thomas S. Kuhn, *The Structure of Scientific Revolutions,* 2d ed. (Chicago: University of Chicago Press, 1970), which in turn was informed by Ludwig Fleck, *Genesis and Development of a Scientific Fact* (Chicago: University of Chicago Press, 1979). A list of books and articles in this field that I consider to be significant includes the following: Pnina, Abir-Am, "How Scientists View Their Heroes: Some Remarks on the Mechanism of Myth Construction," *Journal of the History of Biology* 15 (1982), 281–315; Barry Barnes and David Bloor, "Relativism, Rationalism and the Sociology of Knowledge," in *Rationality and Relativism,* ed. M. Hollis and S. Lukes (Oxford: Blackwell, 1982); Barry Barnes and Steven Shapin, eds., *Natural Order: Historical Studies of Scientific Culture* (Beverly Hills, Calif.: Sage Publications, 1979); Wiebe Bijker, Thomas Hughes, and Trevor Pinch, eds., *The Social Construction of Technological Systems: New Directions in the Sociology and History of Technology* (Cambridge, Mass.: MIT Press, 1988); Alberto Cambrosio and Peter Keating, "Going Monoclonal: Art, Science, and Magic in the Day to Day Use of Hybridoma Technology," *Social Problems* (forthcoming); Harry M. Collins,

Changing Order: Replication and Induction in Scientific Practice (Beverly Hills, Calif.: Sage Publications, 1985); Boelie Elzen, "The Ultracentrifuge," *Social Studies of Science* (forthcoming); Randall Collins and Sal Restivo, "Robber Barons and Politicians in Mathematics: A Conflict Model of Science," *Canadian Journal of Sociology* 8, no. 2 (1983), 199–227; Joan H. Fujimura, "The Construction of Doable Problems in Cancer Research," unpublished ms., Tremont Research Institute, 1986; Peter Galison, *How Experiments End* (Chicago: University of Chicago Press, 1987); Gerald Holton, *The Scientific Imagination: Case Studies* (Cambridge: Cambridge University Press, 1978), and *Thematic Origins of Scientific Thought: Kepler to Einstein* (Cambridge, Mass.: Harvard University Press, 1973); Karin Knorr-Cetina, *The Manufacture of Knowledge: An Essay on the Constructivist and Contextual Nature of Science* (Oxford and New York: Pergamon, 1981); Karin Knorr-Cetina, Roger Krohn, and R. D. Whitley, eds., *The Social Process of Scientific Investigation* (Dordrecht: Reidel, 1981); Karin Knorr-Cetina and Michael Mulkay, eds., *Science Observed: Perspectives on the Social Study of Science* (London: Sage Publications, 1983); Bruno Latour, *Science in Action* (Cambridge, Mass.: Harvard University Press, 1987); Bruno Latour and Steve Woolgar, *Laboratory Life: The Social Construction of Scientific Facts* (Beverly Hills, Calif.: Sage Publications, 1979); Michael Lynch, *Art and Artifact in Laboratory Science* (London: Routledge and Kegan Paul, 1985); D. MacKenzie, "Statistical Theory and Social Interests: A Case Study," *Social Studies of Science* 8 (1978), 35–83; Andy Pickering, *Constructing Quarks: A Sociological History of Particle Physics* (Chicago: University of Chicago Press, 1984); Trevor Pinch, *Confronting Nature* (Dordrecht: Reidel, 1986); Sal Restivo, *The Social Relations of Physics, Mysticism, and Mathematics* (Dordrecht: Reidel, 1983); and "Commentary: Some Perspectives in Contemporary Sociology of Science," *Science, Technology, & Human Values* 6 (1981), 22–30; B. Wynne, "C. G. Barkla and the J Phenomenon: A Case Study of the Treatment of Deviance in Physics," *Social Studies of Science* 6 (1976), 307–347.

1. Touring the Site: Powerful Places in the Laboratory

1. For a discussion of environmental pressures on the organization of human groups, see Robin Fox, *Kinship and Marriage: An Anthropological Perspective* (Harmondsworth, England: Penguin, 1976), and more specifically, E. E. Evans-Pritchard, *The Nuer: A Description of the Modes of Livelihood and Political Institutions of a Nilotic People* (New York: Oxford University Press, 1978).

2. See Stanford University *Campus Reports*, November 11, 1974, p. 11.

3. Douglas William Dupen, *The Story of Stanford's Two Mile Long Accelerator*, SLAC Report no. 62 (Stanford: Stanford Linear Accelerator Center, May 1966).

4. For an explication of the relation of the signified and their signifiers, see Roland Barthes, *Elements of Semiology*, trans. Annette Lavers and Colin Smith (New York: Hill and Wang, 1977).

5. For a report on the incident, see *The Beam Line* (Stanford Linear Accelerator Center), December 20, 1971. This in-house newsletter has been published under a variety of names over the years; I always use the title as given in the issue cited.

6. For a history of these coming-to-power politics, see Daniel J. Kevles, *The Physicists: The History of a Scientific Community in Modern America* (New York; Alfred A. Knopf, 1978; Cambridge, Mass.: Harvard University Press, 1987).

7. For a profile of the cafeteria manager and the ambience she creates, see *The Beam Line*, December 20, 1971. The American cultural connotations of kitchens are analyzed by Nan Bauer Maglin in "Kitchen Dramas," and Joan Greenbaum in "Kitchen Culture/Kitchen Dialectic," both in *Heresies*, vol. 11, no. 3 (1981), pp. 42–46, 59–61.

8. "The systematic study of spatial factors in face to face groups" is the focus of Robert Sommer, *Personal Space: The Behavioral Basis of Design* (Englewood Cliffs, N.J.: Prentice Hall, 1969), which includes a discussion of seating arrangements in public places. His work, however, concentrates on ephemeral groups, not those composed of people familiar with one another over significant periods of time. Familiarity in public places is studied by Sherri Cavan in "Bar Sociability" and, to a lesser degree, by William Foote Whyte in "The Social Structure of the Restaurant." Both studies appear in *People in Places: The Sociology of the Familiar*, ed. Arnold Birenbaum and Edward Sagarin (New York: Praeger, 1973), pp. 143–154, 244–256.

9. The role of clothing in social identity has recently received attention in Alison Lurie, *The Language of Clothes* (New York: Random House, 1980), and John Molloy, *Dress for Success* (New York: Warner, 1976), and *The Woman's Dress for Success Book* (New York: Warner, 1977). These works address clothing as a sign of culture, region, status, and class, which they quite properly define in terms of the meaning of details.

10. Molloy's *Woman's Dress for Success Book* (n. 9) has had many imitators and much attention, but all of these studies deal with the business world. The subtle and severe dress codes for women in academia and laboratories have received no attention.

11. "Campus once domain of sea creatures," *San Jose Mercury*, January 4, 1978 (reprinted in *SLAC Beam Line*, March 1978, p. 4).

12. According to B. L. Harrigan, "reference to military rank, titles, and actions have a predominantly pompous tone. They relate closely to the hierarchical structure and identify business as a glorious wartime battle." She includes a glossary of military terms, noting the symbolic connotations their use in non-military contexts invokes. See her *Games Mother Never Taught You: Corporate Gamesmanship for Women* (New York: Warner, 1977), pp. 99–104.

13. This topic is analyzed in terms of public urban spaces by Kevin Lynch in his *What Time Is This Place* (Cambridge, Mass.: MIT Press, 1972), particularly in his discussion of "episodic contrast" (pp. 173–184). For an investigation of the social control implicit in the scheduling of recurrent behavior, see Barry Schwartz, "Notes on the Sociology of Sleep" in *People in Places*

(n. 8), and Herbert Marcuse, *Eros and Civilization: A Philosophical Inquiry into Freud* (Boston: Beacon, 1956), who argues that controlling "the flux of time is society's most natural ally in maintaining law and order" (p. 231).

14. In 1978, 222 women were employed at SLAC. Five of them, all Caucasian, were fully fledged, nonstudent particle physicists, but four of these five were visitors. Of the nearly 1,100 men employed, 91 were nonstudent particle physicists, of whom ten were visitors. Five of the male physicists were Asian-American; one was American Indian.

15. See reports in the *SLAC Beam Line* in March 1975, April 1975, May 1975, October 1976, August 1977, November–December 1977, and February 1978.

16. For studies of the impact of kinesic behavior, as shaped by the built environment, on social action, see Kent C. Bloomer and Charles W. Moore, *Body, Memory, and Architecture* (New Haven: Yale University Press, 1977), and Kevin Lynch, *The Image of the City* (Cambridge, Mass.: MIT Press, 1960). For a survey of the study of the use and meaning of space, see Emrys Jones and John Eyles, *An Introduction to Social Geography* (Oxford: Oxford University Press, 1977), pp. 5–63.

17. The contrast between outward conformity and the cultural value placed on individualism has been noted by many commentators on American culture. See, for example, Alexis de Tocqueville, *Democracy in America*, ed. Richard D. Herrner (New York: Mentor, 1961): he argues that Americans are inclined by their principle of political equality to establish "small private societies, united together by similitude of conditions, habits, and manners" (p. 247). Sacvan Bercovitch in *The American Jeremiah* (Madison: University of Wisconsin Press, 1978), discussing American writers of the nineteenth century, claims that for these writers "identification with America as it ought to be impels the writer to withdraw from what is in America." Paradoxically, "the ideals that prompt [their] isolation enlist individualism itself, aesthetically, morally, and mythically, into the service of society" (pp. 181–182). For a study which stresses the American capacity for conformity in the twentieth century, see David Riesman, *The Lonely Crowd: A Study of the Changing American Character* (New Haven: Yale University Press, 1950), and William H. Whyte, Jr., *The Organization Man* (New York: Simon and Schuster, 1956).

18. See Dupen, *Story of Stanford's Accelerator* (n. 3), pp. 59–65.

19. Further information about the power supply system for the accelerator can be found in Dupen, *Story of Stanford's Accelerator* (n. 3), pp. 40–45, 91–97; *1970 Annual Report of the Stanford Linear Accelerator Center* (Stanford: United States Atomic Energy Commission, n.d.), pp. 20–26; and *Stanford Linear Acccelerator Center 1971 Annual Report* (Stanford: U.S. Atomic Energy Commission, n.d.), pp. 39–41. All three of these publications have bibliographies with references to the scientific literature produced at SLAC on the entire accelerator system.

20. On the construction of the accelerator, see Dupen, *Story of Stanford's Accelerator* (n. 3), pp. 82–88; *1970 Annual Report* (op. cit., n. 18), pp. 4–7.

21. On the injector, acceleration process, and focusing, see Dupen, *Story of Stanford's Accelerator* (n. 3), pp. 28–58, 89–91, 98–106; *1970 Annual Report*

(n. 18); pp. 27–46. See also Richard B. Neal, "SLAC: The Accelerator," *Physics Today*, April 1967, pp. 27–41.

22. Dupen, *Story of Stanford's Accelerator* (n. 3), pp. 70–81, 66–69, 125.

23. Ibid., pp. 53–58, 115. A history of this laboratory is being written by Bruce Wheaton of the University of California at Berkeley. Three laboratory officials have extensive personal collections of documents.

24. *Stanford Linear Accelerator Center 1971 Annual Report* (n. 18), pp. 42–43; "Taming SLAC's Beam," *The Beam Line*, April 23, 1973; Warren Struven, "Computer Control of the SLAC Accelerator," *SLAC Beam Line*, May 20, 1975.

25. "SLAC Plans Anger Neighbors," *The Stanford Daily*, April 22, 1981, p. 3.

26. See Dupen, *Story of Stanford's Accelerator* (n. 3), p. 62.

27. For a discussion of these issues, see Mary Douglas, *Purity and Danger: An Analysis of the Concepts of Pollution and Taboo* (London: Routledge and Kegan Paul, 1978), pp. 55, 115, and passim; and Mary Douglas and Aaron Wildavsky, *Risk and Culture: An Essay on the Selection of Technological and Environmental Dangers* (Berkeley: University of California Press, 1982). See also Bruno Bettelheim, *Symbolic Wounds: Puberty Rites and the Envious Male* (New York: Collier Books, 1971); Edmund Leach, "Magical Hair," *Journal of the Royal Anthropological Institute*, vol. 88 (1958), pp. 147–164; and Raymond Firth, *Symbols: Public and Private* (Ithaca, N.Y.: Cornell University Press, 1975), pp. 262–298.

28. Mary Douglas, *Purity and Danger* (n. 26), pp. 2–5.

29. For a description of the facilities at Tsukuba, see Henry Birnbaum, "Japan Builds a Science City," *Physics Today*, February 1975, pp. 42–48, and "Tsukuba Science City," in *Guide to World Science*, vol. 17: *Japan*, ed. D. B. Forbes (Guernsey, U.K.: Francis Hodgson), pp. 141–147.

30. Personal communication, Thomas Rohlen. For a discussion of the spartan ethos of Japanese elite secondary schools, see Donald Roden, *Schooldays in Imperial Japan: A Study in the Culture of a Student Elite* (Berkeley: University of California Press, 1980).

31. The Japanese novelist Junichiro Tanizaki explores the contrasting values implicit in stark light and subtle shadows in an essay entitled *In Praise of Shadows*, trans. Thomas J. Harper and Edward G. Seidensticker (New Haven: Leete's Island Books, 1977).

32. Kiyonobu Itakura and Eri Yagi, "The Japanese research system and the establishment of the Institute of Physical and Chemical Research"; Hiroshige Tetu, "Social conditions of prewar Japanese research in nuclear physics"; and Yoshinori Kaneseki, "The elementary particle theory group"; all in Shigeru Nakayama, David L. Swain, Eri Yagi, eds., *Science and Society in Modern Japan: Selected Historical Sources* (Tokyo: University of Tokyo Press, and Cambridge, Mass.: MIT Press, 1974).

33. For a description of KEK specifications, see Tetsuji Nishikawa, "The Japanese 12 GeV Accelerator," KEK Preprint 2 (1974), and KEK Annual Reports for 1974, 1976, 1977, and 1980, National Laboratory for High Energy Physics, Oho-machi, Tsukuba-gun, Ibaraki-ken, 300–32, Japan.

34. Liza Crihfield Dalby, personal communication. See her anthropological study, *Geisha* (Berkeley: University of California Press, 1983).

35. Lillian Hoddeson, "Establishing KEK in Japan and Fermilab in the United States: Internationalism, Nationalism, and High Energy Accelerators," *Social Studies of Science*, April 1983, p. 3.

2. Inventing Machines That Discover Nature: Detectors at SLAC and KEK

1. Ian Hacking, "Do We See through a Microscope?" *Representing and Intervening: Introductory Topics in the Philosophy of Natural Science* (Cambridge: Cambridge University Press, 1983).

2. Bruno Latour and Steve Woolgar, *Laboratory Life: The Social Construction of Scientific Facts* (Beverly Hills, Calif: Sage Publications, 1979), pp. 51, 63–69, 242, and 259, n. 15.

3. Donald A. Glaser, "The Bubble Chamber," *Scientific American*, February 1955, pp. 46–50. For another perspective on the development of the bubble chamber and its implications for the organization of work in high energy physics, see Peter Galison, "Bubble Chambers and the Experimental Workplace," in *Observation, Experiment, and Hypothesis in Modern Physical Science*, ed. Peter Achinstein and Owen Hanaway (Cambridge, Mass.: MIT Press, 1985), pp. 353–359.

4. Glaser, "The Bubble Chamber" (n. 3).

5. "'Moratorium' Party for 82–Inch Bubble Chamber," *The Beam Line*, December 4, 1973.

6. Gerald K. O'Neil, "The Spark Chamber," *Scientific American*, August 1962, pp. 36–43; and D. E. Yount, "The Streamer Chamber," *Scientific American*, October 1967, pp. 38–46. For additional information about different types of detectors of this generation, see Georges Charpak, "Multiwire and Drift Proportional Chambers"; William J. Willis, "The Large Spectrometers"; Jack Sandweiss, "The High Resolution Streamer Chamber"; and D. R. Nygren and J. N. Marx, "The Time Projections Chamber"; all in *Physics Today*, October 1978, pp. 28–53.

7. "'Moratorium' Party" (n. 5).

8. "Forty Inch Bubble Chamber—Now World's Fastest," *The SLAC News*, September 1971; "Hybrid 40–inch Bubble Chamber Studies 'Inelastic Muon-Proton Scattering,'" *The Beam Line*, March 7, 1973; "Cerenkov Counter Being Built for New Hybrid Bubble Chamber Facility," *SLAC Beam Line*, December 1974.

9. Gordon Bowden, "Hybrid Bubble Chambers: An Operator's View," *The Beam Line*, February 6, 1973.

10. "Rapid Cycling Bubble Chamber," *The SLAC News*, October 16, 1970; "Rapid Cycling Bubble Chamber TURNS ON," *The Beam Line*, April 5, 1973.

11. "SLAC and MIT Collaboration Studies Proton Structure," *The SLAC News*, February 26, 1970; Steve Kociol, "Does Time Run Backwards?" *The SLAC News*, July 31, 1970; Charles Oxley, "MPC Improves Spectrometer

Performance," *The SLAC News*, June 2, 1971; Bill Kirk, "End Station A Spectrometers," *SLAC Beam Line*, March 20, 1975, pp. 7–11.

12. S. Kociol, "'Vector Dominance' Works—Or Does It?" *The SLAC News*, April 19, 1971; Charles Oxley, "LASS Construction Underway," *The SLAC News*, April 1972.

13. He believed that his sense of good experimental work had been learned in his teachers' labs in Scotland; he was inspired too by the examples of earlier physicists' experimental equipment that had been displayed in the dining hall of his university laboratory. Scotland has an important history of empirical scientific research. This physicist knew that history through those devices in the dining hall and through his teachers' stories. See for example, Richard Olson, *Scottish Philosophy and British Physics, 1750–1880* (Princeton, N.J.: Princeton University Press, 1975). On multiwire proportional chambers, see Charpak, "Multiwire and Drift Proportional Chambers" (n. 6).

14. Bill Kirk, "An Introduction to Colliding Beam Storage Rings," *SLAC Beam Line*, August 1974, pp. 3–6. See also Martin L. Perl and William T. Kirk, "Heavy Leptons," *Scientific American*, March 1978, pp. 50–57. For a brief history of colliding beam facilities by a physicist active in the design of several of them, see John Rees, "The Last of the Rings?" *CERN Courier International Journal of High Energy Physics*, July–August 1986, pp. 1–3.

15. Bill Kirk, "The Positron Source Job," *SLAC Beam Line*, February 1975, pp. 5–8.

16. Burton Richter, "Burton Richter: A Scientific Autobiography," *SLAC Beam Line*, November 1976, p. 8.

17. "LBL, SLAC Designing Colliding Beam Accelerator," *The Beam Line*, June 20, 1972, p. 4.

18. J.-E. Augustin et al., "Discovery of a Narrow Resonance in e+e− Annihilation," *Physical Review Letters*, December 2, 1974, pp. 1406–1408.

19. J. D. Bjorken, "The 1976 Nobel Prize in Physics," *Science*, 19 November 1976, pp. 825–826, 865–866.

20. Bill Kirk, "Preprints, Publications, and All That," *SLAC Beam Line*, July 1977, pp. 5–8.

21. Martin Einhorn and Chris Quigg, "On the New Narrow Resonances," *SLAC Beam Line*, March 20, 1975, p. 6.

22. For a description of KEK specifications, see *KEK Annual Report 1976*, National Laboratory for High Energy Physics, Oho-machi, Tsukuba-gun, Ibaraki-ken, Japan.

23. In my study of style in detectors I am indebted to the analytic, anthropological work of the archeologist Heather Lechtman. See, especially, "Style in Technology—Some Early Thoughts" in *Material Culture: Styles, Organization, and Dynamics of Technology*, ed. Heather Lechtman and Robert S. Merrill (St. Paul, Minn.: West Publishing Co. 1977; 1975 Proceedings of the American Ethnological Society), and "Andean Value Systems and the Development of Prehistoric Metallurgy," *Technology and Culture*, January 1984.

Walter Vincenti has made a major contribution to my understanding of the relation between tacit, prescriptive, and descriptive knowledge in science and technology in his analysis of innovation, design, and production in engineering in his "Technical Knowledge without Science: The Innovation of Flush Riveting in American Airplanes, c.1930–c.1950," *Technology and Culture*, July 1984, pp. 540–546. On the ethnographic observation and analysis of engineering practices I have been influenced strongly by L. L. Bucciarelli of MIT. See his "Engineering Design Process," in *Making Time: Ethnographies of High-Technology Organizations*, ed. Frank Dubinskas (Philadelphia: Temple University Press, 1988), pp. 92–122.

3. Pilgrim's Progress: Male Tales Told During a Life in Physics

1. Clifford Geertz, "From the Natives' Point of View: On the Nature of Anthropological Understanding," in *Meaning in Anthropology*, ed. Keith H. Basso and Henry A. Selby (Albuquerque, N.M.: University of New Mexico Press, 1976), p. 235.

2. For a study of another culture's organization of emotional states in boys and men during a very long socialization process, see Gilbert H. Herdt, *Guardian of the Flute: Idioms of Masculinity* (New York: McGraw-Hill, 1981). For an analysis of the "psychological adjustments" appropriate to different career stages in America, see Gene W. Dalton, Paul H. Thompson, and Raymond L. Price, "The Four Stages of Professional Careers: A New Look at Performance by Professionals," *Professional Dynamics*, Summer 1977, pp. 19–42. For a study of how affect is linked "to characteristic sorts of activity that change through the life cycle," see Michelle Zimbalist Rosaldo, *Knowledge and Passion: Ilongot Notions of Self and Social Life* (Cambridge: University of Cambridge Press, 1980), p. 63 and passim. For a reevaluation of the anthropological analysis of emotions, see Renato Rosaldo, "Grieving and the Anthropology of Emotions," in *Text, Play, and Story: The Construction and Reconstruction of Self and Society*, ed. by Stuart Plattner, Proceedings of the American Ethnological Society (Washington, D.C.: American Ethnological Society, 1983). The following nonanalytic, anecdotal accounts of elite male socialization in the United States provide data on the cultural construction of emotion: Scott Turow, *One L: An Inside Account of Life in the First Year at Harvard Law School* (New York: Penguin Books, 1978); Charles LeBaron, *Gentle Violence: An Account of the First Year at Harvard Medical School* (New York: Marek, 1981); and Fran Worden Henry, *Toughing It Out at Harvard* (New York: McGraw-Hill, 1984).

3. On analogic thinking in science, see Mary B. Hesse, *Models and Analogies in Science* (Notre Dame, Ind.: University of Notre Dame Press, 1979).

4. Eyrind H. Wichmann, *Quantum Physics*, Berkeley Physics Course, vol. 4 (New York: McGraw-Hill, 1971). See also Thomas S. Kuhn, "The Essential Tension: Tradition and Innovation in Scientific Research," in *Scientific Creativity: Its Recognition and Development*, ed. C. W. Taylor and Frank Barron (New York: Wiley and Sons, 1963), p. 352.

5. On the rhetoric of images and captions, see Roland Barthes, *Empire*

of Signs, trans. Richard Howard (New York: Hill and Wang, 1982); *Roland Barthes*, trans. Richard Howard (New York: Hill and Wang, 1977); and *Image Music Text*, trans. Stephen Heath (New York: Hill and Wang, 1977).

6. Richard Feynman, *The Feynman Lectures in Physics* (Reading, Mass.: Addison-Wesley, 1964), p. xx.

7. Roland Barthes analyzes the popular image of Einstein as genius in "The Brain of Einstein," *Mythologies*, trans. Annette Lavers (New York: Hill and Wang, 1972), pp. 68–70. The photograph of Einstein I have most often seen in physicists' studies is the poster in which he is riding a bicycle awkwardly.

8. W. K. H. Panofsky and R. H. Dalitz, *Particle Physics* (thirteen chapters reprinted by SLAC, n.d., from *Nuclear Energy Today & Tomorrow*.)

9. For an introduction to the scholarly literature on gender, language, and power see Barrie Thorne and Nancy Henley, eds., *Language and Sex: Difference and Dominance* (Rowley, Mass.: Newbury House, 1975); Sally McConnell-Ginet, Ruth Borker, and Nelly Furman, eds., *Women and Language in Literature and Society* (New York: Praeger Publishers, 1980); and Dale Spender, *Man Made Language* (London: Routledge and Kegan Paul, 1980). For a study of the relation between linguistic competition and social action, see Alan Dundes, Terry Leach, and Bora Ozkok, "The Strategy of Turkish Boys' Dueling Rhymes," in *Directions in Sociolinguistics: The Ethnography of Communication*, ed. John J. Gumperz and Dell Hymes (New York: Holt Rinehart and Winston, 1972), pp. 130–160. I am indebted to the observations of Professor John Law, University of Keele, Professor Barrie Thorne, University of Michigan, and Frank A. Dubinskas, Boston College, for their personal communications of their experiences and observations of the rhetoric of sexual domination in scientific communities, and to Professor S. A. Edwards, University of Pittsburgh, for his observations on how "vulgar," male, sexist discourse serves to silence women in order to gain power over them. For comments on the role of male "sexual vernacular" in the American business community, see Betty Lehan Harragan, *Games Your Mother Never Taught You: Corporate Gamesmanship for Women* (New York: Warner Books, 1977), pp. 110–116, and Rosabeth Moss Kanter, *Men and Women of the Corporation* (New York: Basic Books, 1977), pp. 223–226.

10. Anthony Forge discusses this form of training as it is practiced among the Abelan in "Learning to See in New Guinea," in *Socialization: The Approach from Social Anthropology*, ed. Philip Mayer (London: Tavistock, 1970), pp. 269–291, especially 276–278. This practice of telling novices that what they once learned as truth is now to be revised is associated with long novitiates. For analysis of this practice in complex societies, see Rue Bucher and Joan G. Stelling, *Becoming Professional*, Sage Library of Social Research, vol. 46 (Beverly Hills, Calif.: Sage Publications, 1977), and Charles L. Bosk, *Forgive and Remember: Managing Medical Failure* (Chicago: University of Chicago Press, 1979), and Benson R. Snyder, *The Hidden Curriculum* (New York: Knopf, 1971).

11. Stephen Brush, "Should the History of Science Be Rated X?" *Science*, 22 March 1974, p. 1170.

12. Francis Bacon, *The New Organon and Related Writings*, ed. Warhat,

pp. 352, 358. P. B. Medawar also makes this point in his *Advice to a Young Scientist* (New York: Harper and Row, 1979).

13. Stephen Brush, "Should the History of Science Be Rated X?" (n. 11), p. 1164.

14. Robert K. Merton, "Priorities in Scientific Discovery: A Chapter in the Sociology of Science," *American Sociological Review*, December 1957, pp. 635–659. Following the functionalist social system theory and the multivariate statistical analysis that Merton advocated, American sociologists of science have generated many studies of the "reward system" in science. See Bernard Barber, "Science: The Sociology of Science," *International Encyclopedia of the Social Sciences* (1968), vol. 14, pp. 92–100. Nevertheless, in a review article, Mulkay states that the "lack of direct data on commitment on the 'social norms of research' is astonishing, given that these norms were first formulated over 30 years ago, and given the frequency with which they have been mentioned in the literature." M. J. Mulkay, "Sociology of the Scientific Research Community," in *Science, Technology, and Society: A Cross-Disciplinary Perspective*, ed. Ina Spiegel-Rosing and Derek de Solla Price (Beverly Hills, Calif.: Sage Publications, 1977), pp. 135 and 97–99. See also M. J. Mulkay, "Some Aspects of Cultural Growth in the Natural Sciences," *Social Research*, Spring 1969, pp. 22–52.

15. Mulkay, "Sociology of the Scientific Research Community" (n. 14), p. 108.

16. Brush, "Should the History of Science Be Rated X?" (n. 11).

17. Northrop Frye, *Anatomy of Criticism: Four Essays* (Princeton, N.J.: Princeton University Press, 1971), pp. 33–36.

18. Gregory Bateson, "Style, Grace and Information in Primitive Art," in *Steps to an Ecology of Mind* (New York: Ballantine Books, 1972), p. 148.

19. On the invention of the idea of creativity, see Roy Wagner, *The Invention of Culture* (Chicago: University of Chicago Press, 1981), pp. 140–145. See also Raymond Williams, *Keywords: A Vocabulary of Culture and Society* (New York: Oxford University Press, 1976), pp. 72–74.

20. See n. 2, above.

21. By referring to this relationship as avuncular, I am invoking the anthropological notion of the "avunculate": "the right and duties the maternal uncle has toward his sisters' sons and his power over them," together with the informal and ritual behavior appropriate to that relationship. Robin Fox, *Kinship and Marriage: An Anthropological Perspective* (Harmondsworth, England: Penguin Books, 1976), p. 105. See also A. R. Radcliffe Brown, "On Joking Relationships" and "A Further Note on Joking Relationships," *Structure and Function in Primitive Society* (New York: The Free Press, 1965), pp. 90–132, and Gregory Bateson, *Naven: A Survey of the Problems Suggested by a Composite Picture of the Culture of a New Guinea Tribe Drawn from Three Points of View* (Stanford, Calif.: Stanford University Press, 1958), pp. 35–53, 74–85, and passim.

22. I have no data on female advisors.

23. *Roland Barthes* (n. 5), p. 99.

24. The sample included approximately 30 letters of recommendation

written to one group leader, in which I noted 92 qualitative, evaluative remarks, five of which I saw repeated in at least 10 letters, and two were repeated six times: good judgment (10); careful, meticulous, thorough (13); delivered (10); hard and willing worker (16); good colleague (11); good physicist (10); independent (6); intelligent (6). One was called original, two were said to be able to express themselves cogently, and two were identified as ready for the responsibility to organize a project.

25. For a discussion of the role of error in learning, see Gregory Bateson, *Naven* (n. 21), p. 274.

26. For a study of the "crucial social support mechanisms" in science, see Ian I. Mitroff, Theodore Jacob, and Eileen Trauth Moore, "On the Shoulders of the Spouses of Scientists," *Social Studies of Science*, August, 1977, pp. 303–327.

27. Northrop Frye, *Anatomy of Criticism* (n. 17), pp. 33–38.

28. Ibid.

29. Nostalgia is a form of historical consciousness. According to Dean MacCannell, *The Tourist: A New Theory of the Leisure Class* (New York: Schocken Books, 1976), p. 3, nostalgia is a key component in the ethos of progress: "The progress of modernity . . . depends on its very sense of instability and inauthenticity. For moderns, reality and authenticity are thought to be elsewhere: in other historical periods and other cultures, in purer, simpler lifestyles. In other words, the concern of moderns for 'naturalness,' their nostalgia and their search for authenticity are not merely casual and somewhat decadent, though harmless, attachments to the souvenirs of destroyed cultures and dead epochs. They are also components of the conquering spirit of modernity—the grounds of its unifying consciousness." Paul Fussell, *The Great War in Modern Memory* (London: Oxford University Press, 1975), has argued that individual memories of the past are shaped by socially constructed, collective narratives which need have little relationship to the once lived experience. The narratives about detectors are success stories.

30. On the negotiation of reputation in American culture, see Erving Goffman, *The Presentation of Self in Everyday Life* (Garden City, N.Y.: Doubleday Anchor Books, 1959).

31. Professor Donald T. Campbell, Syracuse University, personal communication, April 14, 1978.

32. Stanford University, *Campus Report*, October 15, 1980, p. 4; emphasis mine.

33. Gregory Bateson, *Naven* (n. 21), p. 278.

34. For a discussion of the ideology of "entitlement" among the American upper classes, see Robert Coles, *Privileged Ones* (New York: Little, Brown, 1977), pp. 361–409. Roy M. MacLeod surveys the studies of scientific elites in "Changing Perspectives in the Social History of Science," in *Science, Technology, and Society* (n. 14), pp. 106–122.

35. I am indebted to Professor S.A. Edwards, University of Pittsburgh, for this point.

36. Northrop Frye, *Anatomy of Criticism* (n. 17), pp. 39–45.

37. For one formulation of why science need not concern itself with

motives, see Kenneth Burke, *A Grammar of Motives* (Berkeley: University of California Press, 1969), pp. 505–507.

38. Bateson, *Steps to an Ecology of Mind* (n. 18).

39. Fox, *Kinship and Marriage* (n. 21).

40. On the problems of recruitment and job placement in physics, see Martin L. Perl and Roland H. Good, "Graduate Education and the Future of Physics," Eugen Merzbacher, "Physics as a Career in the Seventies," and L. Grodzins, "Where Have All the Physicists Gone?" papers presented at the Conference on Tradition and Change in Physics Graduate Education, Pennsylvania State University, August 1974; "Tradition and Change in Physics Graduate Education," *Physics Today*, November 1974, p. 91; "Changing Career Opportunities for Physicists," *Physics Today*, October 1977, pp. 85–86; and Thomas L. Neff, Melvyn J. Shochet, Walter D. Wales, and Jeremiah D. Sullivan, "Report on HEPAP [High Energy Physics Advisory Panel], Subpanel on High Energy Physics Manpower," February 1978.

41. On growth rates of scientific communities, see Derek de Solla Price, *Science Since Babylon* (New Haven: Yale University Press, 1961), pp. 110–116.

42. My fieldwork to date among the Japanese physicists has not included extensive, detailed study of the training process itself. Although I can provide a brief description, my comments on that process are generally confined to discussion of how trained physicists enter the established groups of the Japanese particle physics community. For discussion on the role of universities and research institutes in the development of Japanese science, see Shigeru Nakayama, "The Role Played by Universities in Scientific and Technological Development in Japan," *Cahiers d'Histoire Mondiale*, Special Issue on Society, Science, and Technology in Japan, vol. 9, no. 2 (1965), pp. 340–363; Kiyonobu Itakura and Eri Yagi, "The Japanese Research System and the Establishment of the Institute of Physical and Chemical Research," Hirosige Tetu, "Social Conditions for Prewar Japanese Research in Nuclear Physics," and Yoshinori Kaneseki, "The Elementary Particle Theory Group," all in *Science and Society in Modern Japan: Selected Historical Sources*, ed. Shigeru Nakayama, David L. Swain, and Eri Yagi (Tokyo: Tokyo University Press; and Cambridge, Mass.: MIT Press, 1974).

43. James R. Bartholomew, "Japanese Modernization and the Imperial Universities, 1876–1920," *Journal of Asian Studies*, February 1978.

44. William K. Cummings, "Understanding Behavior in Japan's Academic Marketplace," *Journal of Asian Studies*, February 1975, pp. 313–341; Michiya Shimbori, "The Academic Marketplace in Japan," *Developing Economies*, December 1969, pp. 617–639.

45. Thomas Rohlen, *For Harmony and Strength: Japanese White Collar Organization in Anthropological Perspective* (Berkeley: University of California Press, 1974).

46. Sharon Traweek, "An Anthropologist Studies Physicists at Tsukuba," *Chuokoron*, January 1987, pp. 145–153 (in Japanese); a translation is available from the author.

47. Richard P. Feynman, "The Development of the Space-Time View of Quantum Electrodynamics," *Science*, 12 August 1966, pp. 699–708.

48. "Burton Richter: A Scientific Autobiography," *SLAC Beam Line*, November 1976, pp. 7–8.

49. For a study which characterizes the historical relationship between scientist and nature as that between a dominant male and passive female, see E. F. Keller, "Baconian Science: A Hermaphroditic Birth," *Philosophical Forum*, Spring 1980, pp. 299–308. See also her "Gender and Science," *Psychoanalysis and Contemporary Thought*, vol. 1, no. 3 (1978), pp. 409–433. Both are revised and included in Keller's *Reflections on Gender and Science* (New Haven: Yale University Press, 1985).

50. On the picaresque genre see Frank W. Chandler, *The Literature of Roguery* (New York: Burt Franklin, 1958), vol. 2, pp. 469–549; Stuart Miller, *The Picaresque Novel* (Cleveland: Case Western Reserve University, 1967); Harry Sieber, *The Picaresque* (London: Methuen, 1977) pp. 58–74; and Alexander Blackburn, *The Myth of the Picaro: Continuity and Transformation of the Picaresque Novel, 1554–1954* (Chapel Hill: University of North Carolina Press, 1979).

51. On gender and genre, see Susan Gubar and Sandra M. Gilbert, *The Madwoman in the Attic: The Woman Writer and the Nineteenth Century Literary Imagination* (New Haven: Yale University Press, 1979).

4. Ground States: Distinctions and the Ties That Bind

1. Roger M. Keesing, *Kin Groups and Social Structure* (New York: Holt, Rinehart and Winston, 1975), pp. 42–43, 78–90; emphasis is Keesing's. On the exchange of women as a binding force in and source of kinship networks, see Gail Rubin, "The Traffic in Women: Notes on the Political Economy of Sex," in *Toward an Anthropology of Women*, ed. Rayna R. Reiter (New York: Monthly Review Press, 1975), pp. 157–210, and Claude Levi-Strauss, *The Elementary Structures of Kinship* (Boston: Beacon Press, 1969).

2. The SLAC theory group has a distinct network comparable in size to that of the experimentalists. Once a year it sends a notice to about thirty-five people, all within the United States, announcing the commencement of the selection process for new postdocs and asking for nominations. In 1975 the announcement was sent to colleagues at four national laboratories (Argonne, Los Alamos, Lawrence Berkeley Laboratory, and Fermilab), the Institute for Advanced Study at Princeton, and twenty-eight universities (Brandeis, Caltech, Chicago, Columbia, Cornell, Harvard, Illinois, Indiana, Johns Hopkins, Maryland, Michigan, Minnesota, Pennsylvania, Princeton, Purdue, Rochester, Rockefeller, Stanford, SUNY Stony Brook, UC Berkeley, UC Irvine, UC Los Angeles, UC Santa Barbara, UC Santa Cruz, US San Diego, Washington, Wisconsin, and Yale). This list is rather exclusive; this memo is not sent to institutions, but to an individual theorist at each of those institutions.

3. It can be argued that Fermilab is the product of a network, organized as a consortium then called Midwestern University Research Associates

(MURA). For a history of the founding of Fermilab, see Lillian Hoddeson, "Establishing KEK in Japan and Fermilab in the United States: Internationalism, Nationalism, and High Energy Accelerators," *Social Studies of Science*, April 1983, p. 3; Anton J. Jachim, *Science Policy Making in the United States and the Batavia Accelerator* (Carbondale, Ill.: Southern Illinois University Press, 1975); Daniel S. Greenberg, *The Politics of Pure Science: An Inquiry into the Relationship between Science and Government* (New York: New American Library, 1967), pp. 207–242.

4. On the rise (and fall) of national scientific communities, see Joseph Ben-David, *The Scientist's Role in Society: A Comparative Study* (Englewood Cliffs, N.J.: Prentice-Hall, 1971); George Basalla, "The Spread of Western Science," in *Comparative Studies in Science and Society*, ed. Sal P. Restivo and Christopher K. Vanderpool (Columbus, Ohio: Charles E. Merrill, 1974), pp. 359–381; and Ina Spiegel-Rosing and Derek de Solla Price, eds., *Science, Technology, and Sociology: A Cross-Disciplinary Perspective* (Beverly Hills, Calif.: Sage Publications, 1977).

5. On moieties, see Robin Fox, *Kinship and Marriage: An Anthropological Perspective* (Harmondsworth, England: Penguin Books, 1976); Roger M. Keesing, *Kin Groups and Social Structure* (n. 1); and J. A. Barnes, *Three Styles in the Study of Kinship* (Berkeley: University of California Press, 1971). For a study of how the understanding of moieties in "traditional" kinship systems might be applied to "complex" societies, see George Peter Murdock, "Political Moieties," in his *Culture and Society* (Pittsburgh: University of Pittsburgh Press, 1965), pp. 332–350.

6. See Chapter 3 on the role of families in the particle physics community.

7. On gender segregation in occupations, see Martha Blaxall and Barbara Reagon, eds., *Women and the Workplace: The Implications of Occupational Segregation* (Chicago: University of Chicago Press, 1976); Cynthia Fuchs Epstein, *Woman's Place: Options and Limits in Professional Careers* (Berkeley: University of California Press, 1970); Florence Howe, ed., *Women and the Power to Change* (New York: McGraw-Hill, 1975); Ruth B. Knudsin, ed., *Women and Success: The Anatomy of Achievement* (New York: William Morrow, 1974); Sara Ruddick and Pamela Daniels, eds., *Working It Out* (New York: Pantheon, 1977); Joyce Lebra, Joy Paulson, and Elizabeth Powers, eds., *Women in Changing Japan* (Stanford: Stanford University Press, 1976); Dorothy Atkinson, Alexander Dallin, and Gail Warshofsky Lapidus, eds., *Women in Russia* (Stanford: Stanford University Press, 1977); William Mandel, *Soviet Women* (Garden City, N.Y.: Anchor Press, Doubleday, 1975); Jessie Bernard, *Academic Women* (University Park: Pennsylvania State University Press, 1964).

For works specifically on women in physics, see Betsy Ancker-Johnson, "Physicist," in *Women and Success*, ed. Knudsin (op. cit., above), pp. 44–49; and Evelyn Fox Keller, "The Anomaly of a Woman in Physics," in *Working It Out*, ed. Ruddick and Daniels (op. cit., above), pp. 77–91. On women in science, see Michele L. Aldrich, "Review Essay: Women in Science," pp. 126–135; Sally Gregory Kohlstedt, "In from the Periphery: American

Women in Science, 1830–1880," pp. 81–96; and Margaret W. Rossiter, "Sexual Segregation in the Sciences: Some Data and a Model," pp. 146–151; all in *Signs*, Special Issue: *Women, Science and Society*, Autumn 1978. See also Betty M. Vetter, "Data on Women in Scientific Research," in *Report on the Participation of Women in Scientific Research*, ed. Janet Walsh Brown, Michele L. Aldrich, and Paula Quick Hall (Washington, D.C.: American Association for the Advancement of Science, March 1978), vol. 2, appendix D.

8. Max Weber, "Science as a Vocation," from *Max Weber: Essays in Sociology*, trans. H. H. Gerth and C. Wright Mills (New York: Oxford University Press, 1958), pp. 129–156; Robert Merton, "Priorities in Scientific Discovery: A Chapter in the Sociology of Science," *American Sociological Review*, December 1957, pp. 635–659.

9. Moses Maimonides, *The Guide of the Perplexed*, trans. Shlomo Pines (Chicago: University of Chicago Press, 1969), pp. 175–176 (italics in the original). On the place of widely disseminated printed information in the development of science, see Elizabeth Eisenstein, *The Printing Press as an Agent of Change: Communications and Cultural Transformations in Early Modern Europe*, 2 vols. (New York: Cambridge University Press, 1979).

10. On preprints in particle physics, see Louise Addis, "SLAC Library Monitors Underground Physics Press," *The SLAC News*, June 2, 1971, pp. 2–3; Bill Kirk, "Preprints, Publications, and All That," *SLAC Beam Line*, July 1977, pp. 5–8; Bill Kirk, "1974–1977 Anti-Preprint Cumulation," *SLAC Beam Line*, May 1978, p. 4; Miles A. Libbey and Gerald Zaltman, "The Role and Distribution of Written Informal Communication in Theoretical High Energy Physics," American Institute of Physics Report no. AIP3DD-1, USAEC Report No. NYO-3732, June 15, 1967. For other studies of written communication among scientists, see Norman Kaplan and Norman Storer, "Science: Scientific Communication," *International Encyclopedia of the Social Sciences* (1968), vol. 14, pp. 112–117; Warren O. Hagstrom, *The Scientific Community* (Carbondale, Ill.: Southern Illinois University Press, 1965); and D. Crane, *Invisible Colleges* (Chicago: University of Chicago Press, 1972).

11. For an excellent analysis of the anthropological literature on gossip, see Sally E. Merry, "Toward a General Theory of Gossip and Scandal," in *Toward a General Theory of Social Control*, ed. Donald Black (New York: Academic Press, 1982); see also the special issue on gossip as social communication in the *Journal of Communication*, Winter 1977.

5. Buying Time and Taking Space: Negotiations, Collaboration, and Change

1. See Norman W. Storer, "The Internationality of Science," pp. 444–460, and Sal P. Restivo and Christopher K. Vanderpool, "The Third Culture of Science," pp. 461–472, in *Comparative Studies in Science and Society*, ed. Sal P. Restivo and Christopher K. Vanderpool (Columbus, Ohio: Charles E. Merrill, 1974).

2. For the role of universities and research institutes in the development of Japanese science, see Shigeru Nakayama, "The Role Played by Universities

in Scientific and Technological Development in Japan," *Cahiers d'Histoire Mondiale*, Special Issue on Society, Science, and Technology in Japan, vol. 9, no. 2 (1965), pp. 340–363; Kiyonobu Itakura and Eri Yagi, "The Japanese Research System and the Establishment of the Institute of Physical and Chemical Research," pp. 158–201, Hiroshige Tetu, "Social Conditions for Prewar Japanese Research in Nuclear Physics," pp. 202–220, and Yoshinori Kaneseki, "The Elementary Particle Theory Group," pp. 221–252, all in *Science and Society in Modern Japan: Selected Historical Sources*, ed. Shigeru Nakayama, David L. Swain, and Eri Yagi (Tokyo: University of Tokyo Press, and Boston, Mass: MIT Press, 1974).

3. For a description of the facilities and research at INS at this time, see the *Institute for Nuclear Study, University of Tokyo, Annual Report, 1975* (Midori-Cho, Tanashi, Tokyo, Japan).

4. Jerry Gaston in his *Originality and Competition in the British High Energy Physics Community* (Chicago: University of Chicago Press, 1973) makes this point about English physicists. I also heard the same point made by French physicists when I visited the laboratories at Orsay in France.

5. See the *1980 KEK Annual Report*, National Laboratory for High Energy Physics, Oho-machi, Tsukuba-gun, Ibaraki-ken, 300–32, Japan, pp. 2, 124.

6. For information on the work at SSRL, see Herm Winick and Bill Kirk, "SSRP: Stanford Synchrotron Radiation Project" (January 1975); Herm Winick, "Synchrotron Radiation Research Meeting" (October 1975); Herm Winick, "An Update on SSRP" (January 1976): Herm Winick, "Synchrotron Radiation (February 1977); H. Winick and Stephen P. Cramer, "Synchrotron Radiation Probes Nitrogen Fixation" (August 1977); all in the *SLAC Beam Line*.

7. On Nishina and the IPCR, see Hiroshige Tetu, "Social Conditions for Prewar Japanese Research," and Kiyonobu Itakura and Eri Yagi, "Japanese Research System" (n. 2).

8. Tetu, "Social Conditions for Prewar Japanese Research" (n. 2), pp. 214–215.

9. For a study of the place in Japan of alliances formed in school, see Donald Roden, *Schooldays in Imperial Japan: A Study in the Culture of a Student Elite* (Berkeley: University of California Press, 1980).

10. For a detailed study of the relationship between KEK and Fermilab, see Lillian Hoddeson, "Establishing KEK in Japan and Fermilab in the United States: Internationalism, Nationalism, and High Energy Accelerators," *Social Studies of Science*, April 1983, pp. 1–48.

11. *Research Institute for Fundamental Physics: 1974* (Kyoto: Research Institute for Fundamental Physics, Kyoto University, 1974).

12. On the difficulty of "reentering" Japanese society after extended periods abroad, see Merry I. White, "The Rites of Return: Re-Entry and Re-Integration of Japanese International Businessmen," paper presented at 31st Annual Meeting, Association for Asian Studies, March 31, 1979, Los Angeles, California.

13. See *Japan in the Miromachi Age*, ed. John Whitney Hall and Toyoda Takeshi (Berkeley: University of California Press, 1977).

14. Shimbori reports that in 1969, 38 percent of university professors in Japan were graduates of Tokyo University and 16 percent from Kyoto University. Michiya Shimbori, "The Academic Marketplace in Japan," *Developing Economies*, December 1969, p. 624.

15. For a historical survey of this network and the special role of Shoichi Sakata at Nagoya, see Kaneseki Yoshinori, "Elementary Particle Theory Group" (n. 2).

16. See Hoddeson, "Establishing KEK and Fermilab" (n. 10), for a discussion of the considerable power exerted by the JPS as a whole (that is, not merely the particle physics community) in the planning and funding of KEK.

17. Chie Nakane, *Japanese Society* (Berkeley: University of California Press, 1970); and Chie Nakane et al., "Japanese Society," *Current Anthropology*, December 1972, pp. 575–582.

18. See Minoru Watanabe, "Japanese Students Abroad and the Acquisition of Scientific and Technical Knowledge," *Cahiers d'Histoire Mondiale*, vol. 9, no. 2 (1965), pp. 254–293; and Gerald Holton, "Striking Gold in Science: Fermi's Group and the Recapture of Italy's Place in Physics," *Minerva*, April 1974, pp. 159–198; and H. J. Jones, *Live Machines: Hired Foreigners and Meiji Japan* (Vancouver: University of British Columbia Press, 1980).

19. On the role of factions in Japanese decisionmaking in one institution, see David Anson Titus, *Palace and Politics in Prewar Japan* (New York: Columbia University Press, 1974); and Ivan P. Hall, "Organizational Paralysis: The Case of Todai," in *Modern Japanese Organization and Decision Making*, ed. E. F. Vogel and S. Vogel (Berkeley: University of California Press, 1975), pp. 304–330.

20. See Takeo Doi, *The Anatomy of Dependence*, trans. John Bester (Tokyo and New York: Kodansha International, 1985), and *The Anatomy of Self: The Individual versus Society*, trans. Mark A. Harbison (Tokyo and New York: Kodansha International, 1986), and Thomas P. Rohlen, "The Promise of Adulthood in Japanese Spiritualism," *Daedalus*, Spring 1976, pp. 125–143.

21. On European physicists' views of the entrepreneurial spirit of American particle physicists, see "CERN Courier's Crystal Ball," *CERN Courier*, December 1977, reprinted in *SLAC Beam Line*, March 1978, p. 8.

22. On Japanese decisionmaking, see Titus, *Palace and Politics in Prewar Japan*, and Hall, "Organizational Paralysis" (n. 19).

23. See Thomas P. Rohlen, *For Harmony and Strength: Japanese White Collar Organization in Anthropological Perspective* (Berkeley: University of California Press, 1974).

24. See Jane M. Bachnik, "Inside and Outside the Japanese Household (*Ie*): A Contextual Approach to Japanese Social Organization," Ph.D. diss., Harvard University, Cambridge, Mass., 1978. See also *Journal of Japanese Studies*, Special Issue on *Ie* Society, Winter 1985.

25. On the transformation of games into organized sport on the model of corporate capitalism in American cultures, see John Dizikes, *Sportsmen and Gamesmen* (Boston: Houghton Mifflin Co., 1981).

26. On the role of the *iemoto* system in the organization of the arts, see Isao Kamakura, "The *Iemoto* System in Japanese Society," *Japan Foundation Newsletter*, October–November 1981, pp. 1–7; on the way an American anthropologist sees contemporary Japanese society as strongly influenced by the *iemoto* system, see Francis L. K. Hsu, *Iemoto: The Heart of Japan* (New York: Schenkman; dist. New York: Wiley, 1975).

27. On core and periphery, see Storer, "Internationality of Science," and Restivo and Vanderpool, "Third Culture of Science" (n. 1).

28. On cultural models of conflict and change, see Victor Turner, *Drama, Fields, and Metaphors* (Ithaca, N.Y.: Cornell University Press, 1974). On conflict and change in Japan, see *Conflict in Japan*, ed. Ellis S. Krauss, Thomas P. Rohlen, and Patricia G. Steinhoff (Honolulu: University of Hawaii Press, 1984), and Junichi Kyogoku, *The Political Dynamics of Japan*, trans. Nobutaka Ike (Tokyo: University of Tokyo Press, 1987).

Epilogue: Knowledge and Passion

1. See Emile Durkheim, *The Elementary Forms of the Religious Life* (New York: Free Press, 1915), and Thomas F. Gieryn, "Durkheim's Sociology of Scientific Knowledge," unpublished manuscript, 1987. The social construction of time and space is, by now, a conventional topic in anthropological ethnographies, the most notable of which is E. E. Evans-Pritchard's study, *The Nuer: A Description of the Modes of Livelihood and Political Institutions of a Nilotic People* (London: Oxford University Press, 1940). More recent studies include P. Bohannan, "Concepts of Time among the Tiv of Nigeria," pp. 315–330, and D. F. Pocock, "The Anthropology of Time Reckoning," pp. 303–314, in *Myth and Cosmos*, ed. J. Middleton (New York: Natural History Press, 1967); E. R. Leach, "Two Essays Concerning the Symbolic Representation of Time," in *Rethinking Anthropology* (London School of Economics Monographs on Social Anthropology, no. 22, 1966); and Robert J. Thornton, *Space, Time and Culture among the Iragu of Tanzania* (New York: Academic Press, 1980). See also, the review article on this subject by J. Goody, "Time: Social Organization," *International Encyclopedia of the Social Sciences* (1968), pp. 30–42. Informative works by nonanthropologists include: P. L. Berger and T. Luckmann, *The Social Construction of Reality* (New York: Doubleday, 1966); H. Marcuse, *Eros and Civilization: A Philosophical Inquiry into Freud* (Boston: Beacon Press, 1956), 231; E. P. Thompson, "Time, Work-Discipline, and Industrial Capitalism," *Past and Present*, no. 38 (1967), pp. 56–97); A. Schutz, *On Phenomenology and Social Relations* (Chicago: University of Chicago Press, 1975); *History and Theory: Studies in the Philosophy of History*, vol. 6 (Special Issue): *History and the Concept of Time* (Middletown, Conn.: Wesleyan University Press, 1966); B. Lang, "Space, Time, and Philosophical Style," *Critical Inquiry*, vol. 2 (1975), pp. 263–280; and H. Meyerhoff, *Time in Literature* (Berkeley: University of California Press, 1968).

2. For an important discussion of the place of representation, reality, and appearance in science as well as the significance of the production of

experiment, as opposed to theory, see Ian Hacking, *Representing and Intervening: Introductory Topics in the Philosophy of Natural Science* (Cambridge: Cambridge University Press, 1983).

3. Harry Collins, "The Seven Sexes: A Study in the Sociology of a Phenomenon, or the Replication of Experiments in Physics," *Sociology*, vol. 9 (1975), pp. 205–224; Bill Harvey, "The Effects of Social Context on the Process of Scientific Investigation: Experimental Tests of Quantum Mechanics," in *The Social Process of Scientific Investigation*, ed. Karin D. Knorr, Roger Krohn, and Richard Whitley (Dordrecht, Holland: D. Reidel, 1981).

4. Heinz Pagels, Plenary Address, Conference on Science, Technology, and Literature, Long Island University, Greenvale, New York, February 1983.

5. Bruno Latour and Steve Woolgar, *Laboratory Life: The Social Construction of Scientific Facts* (Beverly Hills, Calif.: Sage Publications, 1979).

Index